Nanoscale Surface Modification
for Enhanced Biosensing

Guigen Zhang

Nanoscale Surface Modification for Enhanced Biosensing

A Journey Toward Better Glucose Monitoring

 Springer

Guigen Zhang
Department of Bioengineering
Clemson University
Clemson, SC, USA

ISBN 978-3-319-38323-1   ISBN 978-3-319-17479-2   (eBook)
DOI 10.1007/978-3-319-17479-2

Springer Cham Heidelberg New York Dordrecht London

Printed on acid-free paper

Springer International Publishing AG Switzerland is part of Springer Science+Business Media (www.springer.com)

*I would like to dedicate this book to my
family, my wife Sucheng and sons Andrew
and Gordon, for their unconditional support
over the years.*

Preface

By *surface modification*, we often refer to modification of the surface of an implant so that better integration of the implant with surrounding tissues can be achieved. It has rarely been put in the same context as biosensors. In our journey to develop biosensors with enhanced performance by taking advantage of nanotechnology, we learned that surface modification actually plays some crucial roles in affecting the performance of these biosensors. This book provides a whole-spectrum view of the interdependence between surface modification (both physically and chemically) and key parts of a biosensor, including electrode surface area, electrode functionalization, enzyme stability, and mass transport.

In Chap. 1, biosensors are defined and their basic requirements are discussed with respect to other types of sensors. A brief overview of the two major components of biosensors—sensitive elements and signal transduction techniques—is given. For the sensitive elements, the most commonly used ones including antibodies, nucleotides, enzymes, cells, and synthetic molecules are discussed, and the advantages and drawbacks of each are highlighted. For the signal transduction methods, techniques based on mechanical, optical, electromagnetic, electrical, and electrochemical means are reviewed.

The surface area of electrodes in a biosensor plays an important role in affecting the sensor's general performance. For instance, a larger area provides more surface not only for immobilization of sensitive molecules but also for facilitating electron transfer for signal generation. In Chap. 2, with illustration of how the area of planar surfaces can be increased through morphological (physical) modification by forming standing nanopillars on the surface, an aqua-robust way to add nanopillars to planar surfaces is discussed. Furthermore, a recently patented procedure on how to form integrated micro and nano structures on-a-chip is presented for possible biosensor applications.

In Chap. 3, surface functionalization via chemical modification is discussed. Two different functionalization approaches, one using a conducting polymer to entrap enzymes to the surface through its electro-polymerization, and the other using self-assembled monolayer alkanethiols to tether enzymes to the surface, are presented. In each approach, case studies are presented to showcase how the surfaces modified with nanopillars can be optimally functionalized through the tuning of relevant processing parameters. From a sensing-element perspective, two types of molecules are discussed: catalytic enzyme molecules (e.g., glucose oxidase) and affinity-binding molecules (e.g., avidin-biotin couple). Operations of biosensors in terms of target detection, signal measurement, data analysis, and determination of detection sensitivity through calibration are also discussed.

In Chap. 4, the use of gold nanoparticles for improving the stability of enzymes like glucose oxidase for prolonging the functionality of glucose biosensors is explored. The effect of adding nanoparticles in the functionalization process of electrodes with nanopillar-modified surface is evaluated through colloidal characterizations including zeta potential, UV–Vis absorbance, and fluorescence spectroscopy at various molecule–particle coupling stages. Moreover, the sensing performances of electrodes functionalized with addition of nanoparticles are continuously evaluated up to 120 days.

In Chap. 5, a fluidic sensor device with microchannel for fluid transport and target detection by multiple microelectrodes made of standing nanopillars is presented. By examining the effects of flow rate, channel height, and width, case studies are discussed for glucose detection using the developed fluidic sensor device. The effect of nanoparticles is also examined.

Other than providing a whole-field view of the crucial role surface modification plays in enhancing biosensor performance, another unique feature of this book is that discussions are presented in the form of case studies with basic knowledge, such as electrochemistry in terms of amperometry, cyclic voltammetry, impedimetry, and enzymatic kinetics, embedded throughout the text as information in-need or on-demand to make the text easier to follow. Moreover, all the example cases are provided with full technical details and filled with images of the developed structures and results from the characterization of the structures, thus providing a practical guide for readers who might want to develop similar surfaces.

Clemson, SC, USA   Guigen Zhang
January 2015

Acknowledgments

This book summarizes not only the works from my research group, both at the University of Georgia and Clemson University, over the past years, but also the lessons learned first-hand in our efforts to develop enhanced biosensors. The contributions made by my students and associates are crucial part of this journey. Therefore, I would like to acknowledge those who have contributed in a great deal: Venkataramani Anandan, Yeswanth L. Rao, Rajan Gangadharan, Xiaoling Yang, Seung-Jun lee, Andrew Zhang, and Joseph C. Drwiega. Of course, all these works would not be possible without the financial, technical, and administrative support. So my special thanks go to the National Science Foundation, the Bill and Melinda Gates Foundation, the Faculty of Engineering at the University of Georgia, the Department of Bioengineering, and the Institute for Biological Interfaces of Engineering at Clemson University.

Contents

Acronyms

Below is a list of acronyms used through the text.

CPE	Constant phase element
CV	Cyclic voltammetry, cyclic voltammogram
CVD	Chemical vapor deposition
DI	Deionized, as in DI water
DNA	Deoxyribonucleic acid
EIS	Electrochemical impedance spectroscopy
FAD	Flavin adenine dinucleotide
FADH$_2$	Reduced form of flavin adenine dinucleotide
FFT	Fast Fourier transform
GNP	Gold nanoparticle
GOx	Glucose oxidase
MPA	3-Mercaptopropionic acid
MUA	11-Mercaptoundecanoic acid
PBS	Phosphate-buffered saline
PDMS	Polydimethylsiloxane
PVD	Physical vapor deposition
PAA	Porous anodized alumina
PR	Photoresist
RCA	Radio Corporation of America
SAM	Self-assembled monolayers
SERS	Surface enhanced Raman spectroscopy
SEM	Scanning electron microscopy
SPR	Surface plasmon resonance
TEM	Transmission electron microscopy
UV–Vis	Ultraviolet–Visible

Chapter 1
A Brief Overview of Biosensors

Abstract This chapter begins with defining biosensors and their basic requirements by looking at the different challenges a biosensor face in comparison with other sensors. After that, it provides a brief overview of the two major components of biosensors, namely, sensitive elements and the underlying signal transduction techniques. For the sensitive elements, the most commonly used ones including antibodies, nucleotides, enzymes, cells, and synthetic molecules are discussed and the advantages and drawbacks of each are highlighted. For the signal transduction methods, techniques based on mechanical, optical, electromagnetic, electrical, and electrochemical means are reviewed. To close the discussion, it presents an important link between surface modification and the surface properties of electrodes in biosensors.

1.1 What is Biosensor

Biosensors are important devices for identifying and quantifying certain biological species in processes of environmental, food, pharmaceutical, and biomedical concerns. Biosensors generally utilize a biologically sensitive element along with a physical transducer to selectively and quantitatively convert biochemical events into physical signals. The biosensitive element is for target recognition and/or signal generation, and the physical transducer is for signal production or conversion. In many situations, the biological sensitive element consists of biological probes made of molecular species such as antibodies or other proteins, aptamers, or nucleic acids for binding with the target analytes like antigens, ligands, and complementary nucleic acids. In other cases, the sensitive element is made of enzymes for catalyzing certain reactions involving electron exchanges. For the transducer, many physical techniques including mechanical, electrical, optical or electromagnetic, and electrochemical methods are popular choices. In IUPAC (International Union of Pure and Applied Chemistry) definition, a biosensor is "A device that uses specific biochemical reactions mediated by isolated enzymes, immunosystems, tissues,

G. Zhang, *Nanoscale Surface Modification for Enhanced Biosensing*,
DOI 10.1007/978-3-319-17479-2_1

organelles or whole cells to detect chemical compounds usually by electrical, thermal or optical signals" [1]. For example, sensing platforms using the avidin-biotin couple as the biosensitive element along with an electrochemical technique in the form of either voltammetric, impedimetric or amperometric method have been widely utilized to achieve rapid, specific, and sensitive detections [2–5].

The main challenges many biosensors face today include low sensitivity, poor specificity, and proneness to fouling. The advent of nanotechnology presents some promising solutions for alleviating these problems. For example, improvements for the sensitivity and antifouling capability of biosensors have been explored through the modification of the surfaces of biosensor electrodes, namely, the incorporation of nanostructures into the electrode surfaces [6–9]. Nanostructures like gold nanotubes [10] carbon nanotubes [11–13] and gold nanoparticles [14] have also been used to modify the surface of electrodes with improved sensing performance in comparison with conventional unmodified flat electrodes.

1.2   Basic Requirements for Biosensors

A biosensor is, first of all, a sensor. This means that it needs to meet the basic requirements for any sensor: being sensitively responsive and reliable over a long period of time. However, unlike a conventional sensor, a biosensor is often exposed to an environment containing many biological species that are similar in structures and binding behavior. Thus, in addition to meeting the above basic requirements, a biosensor needs to be specific, that is, be responsive only to a specifically targeted analyte species to ensure the usefulness and reliability of the biosensor. Furthermore, because of the harsh and complex biological environment a biosensor often encounters, the loss of sensitivity is a major cause for the compromise of the reliability of a biosensor. This loss is mainly due to either the degradation of the molecular probes or their encapsulation (often termed fouling) by other microorganisms or large molecular weight proteins [15]. Thus for a biosensor, the molecular probes to be used need to have long-lasting activity and antifouling behavior in addition to their specificity. On the transducer side, the underlying signal transduction method should not interfere with the analyte causing false signals. In a nut shell, a biosensor should be sensitive to the target analyte and generate fast responses once detected in the form of reproducible measurements with linear proportionality with analyte concentration and high signal-to-noise ratio, have high selectivity and specificity thus generating response only to target molecules with no interference from non-target molecules, possess sufficient detection limits on both the lower and upper ends, and have sufficient shelf life.

1.3 Various Surface Sensitive Elements

Biosensors can be classified according to the types of their sensitive elements employed. Currently, five types of sensitive elements are mainly being used, namely, antibodies, nucleotides, enzymes, cells, and synthetic molecules [16]. Biosensors using antibodies as the sensitive element operate based on the binding of an antigen to a specific antibody. Biosensors with nucleotides as the sensitive element are usually used to target the genetic materials such as DNA. Biosensors using enzymes as the sensitive element operate based on catalytically induced chemical reactions. The use of enzymes in this class of biosensors adds certain degree of complexity. For instance, while some enzymes require no additional compounds for activity, many enzymes require a cofactor for their activity. Moreover, the catalytic activity of enzymes is governed by the integrity of their native protein conformation. When enzymes are denatured or dissociated, their catalytic activity will be destroyed, which in turn will compromise the reliability of the biosensors. Because of this, this class of biosensors often exhibits a degrading sensing performance over time.

Cell based biosensors are another important class of sensors gaining more and more attention lately. The use of whole cells as the sensitive element is very attractive because cells can provide highly selective and sensitive receptors, channels, and enzymes. The main advantages of cell-based biosensors are that cells have built-in natural selectivity to biologically active chemicals and that cells can react to analytes in a physiologically relevant mode [17, 18]. With a cell-based biosensor, measurements of transmembrane potential, impedance, and metabolic activity can be made. Challenges abound, however, for long-term operations of this class of biosensors because the viability of the cells must be maintained under various harsh operating conditions. To date, cells such as neurons [19], cardiac myocytes [20], liver cells [21], and genetically engineered B cells [22] have been used as the sensitive elements. Besides these cells, microorganisms and bacterial cells have also been used as the sensitive elements in biosensors for the detection and monitoring of environmental pollutants [23] and evaluation of the effectiveness of drugs [24, 25]. Whole cell-based biosensors can offer tremendous benefits for screening drugs and studying the effects of biochemicals on multicellular organisms.

Synthetic molecule based biosensors often use synthetic polymers such as aptamers as the sensitive element [26]. Aptamers are synthetic nucleic acids that can be synthesized to couple (or fit) with amino acids, drugs, proteins, and other non-nucleic molecules. Because of that, this class of biosensors can provide high affinity to a wide array of targets with excellent specificity. Furthermore, these biosensors can maintain prolonged reliability due to the synthetic nature of the polymeric sensitive element which will not denature over time.

To be functional, these sensitive elements need to be attached, or immobilized, to the surface of an electrode by means of surface modification. The electrode is often linked to a underlying transducer mechanism such that the biological recognition, binding or reaction event can be converted into physical signals for quantitative analyses.

1.4   Various Transduction Methods

For the underlying signal transducers, various physical and chemical techniques can be used for converting the biological recognition, binding or reaction events into physical measurable signals. These methods can be generally categorized into mechanical, optical or electromagnetic, electrical, thermal, and electrochemical methods, among others. The main operational principles for the mechanical, optical and electromagnetic, electrical and electrochemical methods are summarized here.

1.4.1   Mechanical Transduction

A mechanical transducer relies on either mechanical deformations or mechanical waves (or acoustic waves) as its sensing mechanism. To implement such a detection method, a mechanical structure in the form of a cantilever beam, a double-clamped beam, or a disc is often used as the underlying transducer with the surface of the transducer modified (or functionalized) by immobilizing a layer of sensitive elements on it for target binding.

In the case of a cantilever beam, a common mode of detection is through the measurement of cantilever deflection caused by either the weight of the bound molecules or the surface stresses generated as a result of molecular binding. The working principle for the former is straightforward. For the latter it relies on the induced differential surface stress produced when molecules bind to one side of the cantilever surface [27–29]. Surface stresses mainly arise from intermolecular forces such as electrostatic interaction or van der Waals. Once generated, the differential stress will cause the cantilever to deflect. According to the classical work by Stoney [30], for a fixed set of cantilever geometric and material properties, its deflection is linearly proportional to the differential surface stress which is related to the amount of molecular binding. The cantilever deflection is often measured by two common techniques. The first one is via an optical means in which a laser beam is focused on the free end of the cantilever and the cantilever deflection is measured with a four-segment photo detector. The second technique is through an electrical means in which a resistive or capacitive circuitry is used to measure the cantilever deflection [31]. This class of mechanical biosensors is capable of detecting mismatches in oligonucleotide hybridization without labeling [32] and of performing protein recognition with extremely high sensitivity. Moreover, this method is compatible with many analyte species in gaseous or aqueous forms [33]. There are limitations as well. If the molecular binding events are exothermic, the heat generated may compromise the detection because a differential thermal stress will also lead to deflection in the cantilever [34]. Another issue is with the nonlinear and viscoelastic nature of the molecular structures which may render it invalid to use Stoneys equations in interpreting the relationship between the measured cantilever deflection and the amount of molecular binding [35, 36].

In the cases of a double-clamped beam or a disc structure, a common mode of transduction is quantifying the changes in the acoustic characteristics such as the resonant frequency, attenuation and phase of wave propagation. In this mode, the mechanical structures operate like oscillators, and a molecular binding event serves as mass loading which often leads to either a shift in the resonant frequency, an increase in amplitude attenuation, or a delay in the phase of wave propagation. Its basic operating principle relies on that any mechanical structure possesses a unique resonate frequency (the lowest Eigen frequency of the structure) along with a certain amount of attenuation and phase of propagation. When molecular binding occurs at the active surface of such a mechanical structure, the mass of the structure and damping to the wave propagation will increase [37]. Under this circumstance, the structure will exhibit certain changes in its wave characteristics when it is perturbed by an external acoustic wave. To increase the detection sensitivity, the mechanical structure (a beam or disc) should possess a high quality factor [38]. In general, the quality factor decreases when the size and damping of the mechanical structure increase. Bulk acoustic waves are more susceptible to liquid-damping induced attenuation than surface acoustic waves, thus detections based on bulk acoustic waves (in the cases of a double-clamped beam or a quartz crystal microbalance) are preferably used in a dry environment and detections based on surface acoustic waves are often used in a liquid environment. A detailed discussion of the applications of bulk and surface acoustic wave devices can be found in reviews by Rao and Zhang [39] and Zhang [40]. By detecting the frequency shift, the attenuation drop, and the phase shift, the amount of bound analyte can be determined. The advantage of this mode of detection is that a single frequency sweep can provide a quick measurement of the mass of the bound molecules at a resolution down to pico-gram level [41]. The challenge for this type of mechanical detection, however, lies in the difficulty in distinguishing the type and the uniformity of the bound species, thus rendering it less specific in biological sensing.

1.4.2 Optical and Electromagnetic Transduction

Optical detection is one of the widely used mechanisms for biosensing because this method can be incorporated into many different types of spectroscopic techniques, including luminescence, absorption, polarization, and fluorescence [42]. With this transduction method, different spectrochemical properties such as amplitude, energy, polarization, decay time, and phase of a target analyte can be measured. These spectroscopic properties can be correlated to the concentration of the analyte of interest.

Of the many optical techniques, fluorescence based detection is probably the most used method. In this method, fluorescent markers that emit light at specific wavelengths are used as detecting labels for the target analytes, and measurements of fluorescent intensity are made for the presence of the targets or the binding of targets to the probes. Many micro-array gene chips use this technique for the detection of hybridization. Furthermore, fluorescence based detection methods have been used to systematically analyze protein–protein and protein–DNA interactions. This technique is capable of single molecule detection [43–45].

The sensing principle based on the evanescent wave is another common mode of optical detection. In this method, an optical waveguide is used to confine the light traveling through the waveguide by total internal reflection. With a major part of the light confined inside the waveguide, a small part of it (i.e., the evanescent wave field) travels through a region that extends about several tens nanometers into the surrounding medium. This evanescent wave can be used for sensing purposes. In a sensing application, the waveguide surface is functionalized with biological sensitive elements, and the change in the optical properties of the evanescent wave is measured in response to the binding of the probe and target molecules. Evanescent wave based biosensors are very selective and sensitive for the detection of low levels of chemicals and biological species and they are suited for the measurement of molecular interactions in-situ and in real time [46]. One of the most used evanescent wave biosensors is the surface plasmon resonance (SPR) biosensor owing to its high sensitivity and simplicity. In an SPR biosensor, the change in the refractive index of the evanescent wave, caused by the interaction between the target molecules and the sensitive probing molecules immobilized on the sensor surface in the evanescent field, is measured.

A well-known electromagnetic detection method is based on the theory of surface-enhanced Raman spectroscopy (SERS). Surface-enhanced Raman scattering is observed for molecules placed close to a rough metal surface featured with silver or gold nanostructures (e.g., nanoparticles or nanowires) because of SPR. This makes SERS a very sensitive detection technique. The working mechanism by which an SERS detection operates is still a debating issue. It is believed that it operates from a local electromagnetic field enhancement provided by an optically active nanoparticle. The electromagnetic effect alone, however, does not account for all that is observed through SERS. Molecular resonances, charge-transfer transitions, and other processes such as ballistic electrons transiently probing the region where the molecule resides and modulating electronic processes of the metal certainly contribute to the rich information that SERS measures [47]. Nevertheless, ultrasensitive analytical strategies and assays based on SERS have been realized [48, 49], in which an enhancement factor as large as 10^{14} is claimed, making it possible for routine detection of Raman signals from single molecules.

1.4.3 Electrical Transduction

Although it has not been as widely used as the mechanical or optical detection methods, electrical detection actually possesses some desirable features as an underlying transducer due to its easy of use. Conductometric and potentiometric techniques are two common modes of electrical transduction, and they mainly rely on the measurement of changes in conductance, impedance, or potential in response to a biological binding event occurring at the electrode surfaces.

Conductometric sensors detect changes in the electrical resistance or impedance between two electrodes [50, 51]. In this case, the changes in resistance or impedance are due to either molecular interactions between nucleotides, proteins, and antigens and antibodies, or excretion of metabolites near the electrode surfaces or in the surrounding media. This mode of detection is attractive because it does not require a specialized reference electrode. So far, this method has been used to detect a wide variety of chemical and biological target species, toxins and nucleic acids, to measure the metabolic activity of microorganisms, and to monitor DNA hybridization [52–54]. Currently, a practical challenge for a conductance base biosensing method is the understanding of the underlying mechanism for the changes in electrical properties of the electrode material caused by molecular adsorption and coupling.

Potentiometric sensors measure the potential changes between electrodes. The most common design of potentiometric sensors uses ion-sensitive field effect transistors or chemical field effect transistors [55]. A pH meter is such an example. Potentiometric sensors have been used to perform detection of hybridization of DNA by measuring the field effect in silicon due to the intrinsic molecular charges on the DNA [56]. Recently, potentiometric sensors have been miniaturized to nanometer dimension through the use of silicon nano-wires [57] and carbon nanotubes [58] for enhanced sensitivity due to the increased surface-to-volume ratio for the electrodes.

1.4.4 Electrochemical Transduction

Biosensors using an electrochemical method as the underlying transducer are often used to measure electron movements resulted from the electrochemical reactions of redox species catalyzed by enzymes typically occurring at a low electrode potential (e.g., from $-1,000$ to $+1,000\,mV$). An electrode potential outside this range could lead to oxygen evolution at the positive end and hydrogen evolution at the negative end. These biosensors are usually configured in a three-electrode format: a working electrode, a counter electrode, and a reference electrode. The reference electrode needs to meet the special requirement of maintaining a constant potential level with respect to the electrolytic solution.

For biological detections, three modes of operations, namely, amperometric, voltammetric, and impedimetric are most commonly used. Amperometric biosensors measure the electrical current generated by the electron exchange between the electrodes and ionic species in response to electrode polarization at a constant potential. The measured steady-state limiting current (due to the encountered diffusion limit) is often linearly proportional to the concentration of the electroactive analyte species. Voltammetric biosensors measure the current-potential relationships (i.e., voltammograms) induced by a redox process. The obtained peak currents and peak potentials (both oxidation and reduction), or limiting currents in the case of sigmoidal voltammograms for nanometer electrodes, are related to the transport phenomena and efficiency as well as the concentration of the redox species. Impedimetric biosensors measure the changes in the complex impedance of an electrochemical process upon cyclic excitations of the working electrode at a predetermined range of frequency. The measured results, often in Bode plots or Nyquist plots, are indicative of the electron transfer resistance which is related to the electrode/solution interfacial properties and the concentration of the analyte.

For the functionalization of these biosensors, enzymes are often used for catalytic reactions and other sensitive receptors (e.g., antibodies, nucleotides, cells and proteins) are used for affinity based detection. In the case of a glucose sensor, the working electrode is usually functionalized with glucose oxidase for catalyzing glucose oxidation, and the current response is measured. Electrochemical based biosensors have been used for the detection of glucose, lactose, urea, lactate, and DNA hybridization [59–64].

1.5   The Surface of Electrodes in Biosensors

Very often a sensitive element and a transduction method are integrated into something we usually call "electrode." In such a situation, the electrode becomes a crucial part of a biosensor: the sensitive element attached on the electrode surface generates a biological or biochemical response and the underlying transduction method converts the biological/biochemical response into a physical signal. The surface property of such an electrode plays a very important role. In terms of sizes, a larger surface area is beneficial not only for the attachment of more sensitive molecules but also for high signal generation or conversion. In terms of materials, since most of these physical transduction methods are built upon inorganic materials, attaching sensitive elements to inorganic materials often requires the use of anchoring molecules [5, 65]. Thus, to improve the performances of biosensors, the modification of electrode surfaces often entails both physical and chemical means.

Physically, morphological changes may be made to the surface. For example, the surface area of electrodes can be increased by modifying the surface with nanostructures because the surface-to-volume ratio of a structure increases as its size decreases [9, 65, 66]. Chemically, the choice of anchoring molecules for attaching sensitive elements to electrodes may affect the sensing performance. To date, many anchoring techniques have been developed using self-assembled monolayer [67–70], conducting polymers [11, 71] and sol–gels [72]. Although surface modification is routine in many biomaterials applications, it is by no means understood how surface modification affects the sensing performance of electrodes in biosensors, especially when the modification encompasses both physical and chemical alterations of the surface properties. In this book we intend to seek an answer to this question by reviewing the things we encountered and lessons learned in our quest to develop biosensors with enhanced performances. We will examine the relevant factors including surface morphological modification and chemical modification in a fluidic environment where mass transport may also play a role.

References

1. http://goldbook.iupac.org/B00663.html, Accessed December 2014
2. Ding, S.J., Chang, B.W., Wu, C.C., Lai, M.F., Chang, H,C.: Impedance spectral studies of self-assembly of alkanethiols with different chain lengths using different immobilization strategies on Au electrodes. Anal. Chim. Acta **554**(1–2), 43–51 (2005)
3. Hou, Y., Helali, S., Zhang, A., Nicole, J.R., Martelet, C., Minic, C., Gorojankina, T., Persuy, M.A., Edith, P.A., Salesse, R., Bessueille, F., Samitier, J., Errachid, A., Akimov, V., Reggiani, L., Pennetta, C., Alfinito, E.: Immobilization of rhodopsin on a self-assembled multilayer and its specific detection by electrochemcial impedance spectroscopy. Biosens. Bioelectron. **21**, 1393–1402 (2006)
4. Hou, Y., Nicole, J.R., Martelet, C., Zhang, A., Jasmina, M.V., Gorojankina, T., Persuy, M.A., Edith, P.A., Salesse, R., Akimov, V., Reggiani, L., Pennetta, C., Alfinito, E., Ruiz, O., Gomila, G., Samitier, J., Errachid, A.: A novel detection strategy for odorant molecules based on controlled bioengineering of rat olfactory receptor I7. Biosens. Bioelectron. **22**, 1550–1555 (2007)
5. Lee, S.J., Anandan, V., Zhang, G.: Electrochemical fabrication and evaluation of highly sensitive nanorod-modified electrodes for a biotin/avidin system. Biosens. Bioelectron. **23**(7), 1117–1124 (2008)
6. Koehne, J., Li, J., Cassell, A.M., Chen, H., Ye, Q., Ng, H.T., Han, J., Meyyappan, M.: The fabrication and electrochemical characterization of carbon nanotube nanoelectrode arrays. J. Mater. Chem. **14**(4), 676–684 (2004)
7. Wang, J., Nosang, M., Minhee, Y., Harold, M.: Glucose oxidase entrapped in polypyrrole on high-surface-area Pt electrodes: a model platform for sensitive electroenzymatic biosensors. J. Electroanal. Chem. **575**, 139–146 (2004)
8. Anandan, V., Rao, Y.L., Zhang, G.: Nanopillar array structures for high performance electro-chemical sensing. Int. J. Nanomed. **1**, 73–79 (2006)
9. Anandan, V., Yang, X., Kim, E., Rao, Y., Zhang, G.: Role of reaction kinetics and mass transport in glucose sensing with nanopillar array electrodes. J. Biol. Eng. **1**, 5 (2007)
10. Delvaux, M., Demoustier-Champagne, S.: Immobilisation of glucose oxidase within metallic nanotubes arrays for application to enzyme biosensors. Biosens. Bioelectron. **18**, 943–951 (2003)
11. Gao, M., Gordon, L.D., Dai, L., Wallace, G.: Biosensor based on vertically aligned carbon nanotubes coated with inherently conducting polymers. Electroanalysis **15**(13), 1089–1094 (2003)
12. Wang, J., Musameh, M.: Carbon nanotube/teflon composite electrochemical sensors and biosensors. Anal. Chem. **75**, 2075–2079 (2003)
13. Wang, J., Mustafa, M.: Carbon nanotube screen-printed electrochemical sensors. Analyst **129**, 1–2 (2004)
14. Bharathi, S., Nogami, M.: A glucose biosensor based on electrodeposited biocomposites of gold nanoparticles and glucose oxidase enzyme. Analyst **126**, 1919–1922 (2001)
15. Ratner, B.D., Hoffman, A.S., Schoen, F.J., Lemons, J.E.: Biomaterials Science, 2nd edn. Academic, Amsterdam (2004)
16. Kubik, T., Bogunia-Kubik, K., Sugisaka, M.: Nanotechnology on duty in medical applications. Curr. Pharm. Biotechnol. **6**, 17–33 (2005)
17. Bousse, L.: Whole cell biosensors. Sensors Actuators B **B34**(1–3), 270–275 (1996)
18. Stenger, D.A., Gross, G.W., Keefer, E.W., Shaffer, K.M., Andreadis, J.D., Ma, W., Pancrazio, J.J.: Detection of physiologically active compounds using cell-based biosensors. Trends Biotech. **19**(8), 304–309 (2001)
19. Borkholder, D.A., Bao, J., Maluf, N.I., Perl, E.R., Kovacs, G.T.A.: Microelectrode arrays for stimulation of neural slice preparations. J. Neurosci. Methods **77**(1), 61–66 (1997)

20. Pancrazio, J.J., Bey, P.P., Cuttino, D.S., Kusel, J.K., Borkholder, D.A., Shaffer, K.M., Kovacs, G.T.A., Stenger, D.A.: Portable cell-based biosensor system for toxin detection. Sensors Actuators B **53**(3), 179–185 (1998)
21. Powers, M.J., Domansky, K., Griffith, L.G.: A microfabricated array bioreactor for perfused 3D liver culture. Biotechnol. Bioeng. **78**(3), 257–269 (2002)
22. Rider, T.H., Petrovick, M.S., Hollis, M.A.: A B-cell based sensor for rapid identification of pathogens, Science **301**, 213–215 (2003)
23. DSouza, S.F.: Microbial biosensors. Biosens. Bioelectron. **16**, 337–353 (2001)
24. Reininger-Mack, A., Thielecke, H., Robitzki, A.A.: 3D-biohybrid systems: applications in drug screening. Trends Biotechol. **20**, 56–61 (2002)
25. Thielecke, H., Mack, A., Robizki, A.: Biohybrid microarrays—impedimetric biosensors with 3D in vitro tissues for toxicological and biomedical screening. Anal. Bioanal. Chem. **369**, 23–29 (2001)
26. Cai, H.L., Lee, H., Hsing, T.M., Ming, I.: Label-free protein recognition using an aptamer-based impedance measurement assay. Sensors Actuators B **114**, 433–437 (2006)
27. Berger, R., Delamarche, E., Lang, H.P., Gerber, C., Gimzewski, J.K., Meyer, E., Guntherodt, H.J.: Surface stress in the self-assembly of alkanethiols on gold. Science **276**, 2021–2023 (1997)
28. Sepaniak, M., Datskos, P., Lavrik, N., Tipple, C.: Microcantilever transducers: a new approach in sensor technology. Anal. Chem. **1**, 568A-575A (2002)
29. Cherian, S., Gupta, R.K., Mullin, B.C., Thundat, T.: Detection of heavy metal ions using protein-functionalized microcantilever sensors. Biosens. Bioelectron. **19**, 41–46 (2003)
30. Stoney, G.: The tension of metallic films deposited by electrolysis. Proc. R. Soc. London Ser. A. **82**(553), 172–175 (1909)
31. Porter, T.L., Eastman, M.P., Macomber, C., Delinger, W.G., Zhine, R.: An embedded polymer piezoresistive microcantilever sensor. Ultramicroscopy **97**, 365–369 (2003)
32. Carrion-Vazquez, M., Oberhauser, A.F., Fowler, S.B., Marszalek, P.E., Broedel, S.E., Clarke, J., Fernandez, M.J.: Mechanical and chemical unfolding of a single protein: a comparison. Biophysics **96**, 3694–3699 (1999)
33. Wu, G.H., Datar, R.H., Hansen, K.M., Thundat, T., Cote, R.J., Majumdar, A.: Bioassay of prostate-specific antigen (PSA) using microcantilevers. Nat. Biotechnol. **19**, 856–860 (2001)
34. Mertens, J., Finota, E., Thundat, T., Fabrea, A., Nadal, M. H., Eyraud, V., Bourillota, E.: Effects of temperature and pressure on microcantilever resonance response. Ultramicro **97**, 119–126 (2003)
35. Zhang, G. Gilbert, J.L.: A new method for real-time and in-situ characterization of the mechanical and material properties of biological tissue constructs. In: Schutte, E., Picciolo, G.L. (eds.) Tissue Engineered Medical Products (TEMPs), ASTM STP 1452, pp. 120–133. ASTM International, West Conshohocken, PA (2004)
36. Zhang, G.: Evaluating the viscoelastic properties of biological tissues in a new way. J. Musculoskelet. Nueronal Interact. **5**(1), 85–90 (2005)
37. Headrick, J.J., Sepaniak, M.J., Lavrik, N.V., Datskos, P.G.: Enhancing chemi-mechanical transduction in microcantilever chemical sensing by surface modification. Ultramicroscopy **97**, 417–424 (2003)
38. Davis, Z.J., Abadal, G., Kuhn, O., Hansen, O., Grey, F., Boisen, A.: Fabrication and characterization of nanoresonating devices for mass detection. J. Vac. Sci. Technol. B **18**(2), 612–616 (2002)
39. Rao, L.R., Zhang, G.: Enhancing the sensitivity of SAW sensors with nanostructures. Curr. Nanosci. **2**(4), 311–318 (2006)
40. Zhang, G.: Nanostructure-enhanced surface acoustic waves biosensor and its computational modeling. J. Sens. **2009**, 11 p (2009). doi:10.1155/2009/215085. Article ID 215085
41. Thundat, T., Wachter, E.A., Sharp, S.L., Warmack, R.J.: Detection of mercury vapor using resonating microcantilevers. Appl. Phys. Lett. **66**, 1695–1697 (1995)
42. Wickline, S.A., Lanza, G.M. Nanotechnology for molecular imaging and target therapy. Circulation **107**, 1092–1095 (2003)

43. Vo-Dinh, T., Cullum, B.: Biosensors and biochips: advances in biological and medical diagnostics. J. Anal. Chem. **366**, 540–551 (2000)
44. Nie, S., Zare, R.N.: Optical detection of single molecules. Anun. Rev. Biophys. Biomol. Struct. **26**, 567–596 (1997)
45. Moerner, W.E., Orrit, M.: Illuminating single molecules in condensed matter. Science **283**, 1670–1676 (1999)
46. Liu, X., Tan, W.: A fiber-optic evanescent wave DNA biosensor based on novel molecular beacons. Anal. Chem. **71**, 5054–5059 (1999)
47. Moskovits, M.: Surface-enhanced raman spectroscopy: a brief retrospective. J. Raman Spectrosc. **36**, 485–496 (2005)
48. Emory, S.R., Haskins, W.E., Nie, S.: Direct observation of size–dependent optical enhancement in single metal nanoparticles. J. Am. Chem. Soc. **120**(31), 8009–8010 (1998)
49. Krug, J.T., Wang, G.D., Emory, S.R., Nie, S.: Efficient raman enhancement and intermittent light emission observed in single gold nanocrystals. J. Am. Chem. Soc. **121**, 9208 (1999)
50. Chen, R.J., Bangsaruntip, S., Dai, H.: Noncovalent functionalization of carbon nanotubes for highly specific elctronic biosensors. Proc. Natl. Acad. Sci. **100**, 4984–4989 (2003)
51. Chen, R.J., Choi, H.C., Dai, H.: An investigation of the mechanism of electronic sensing of protein adsorption on carbon nanotube devices. Am. Chem. Soc. **126**, 1563–1568 (2004)
52. Sosnowski, R.G., Tu, E., Butler, W.F., OConnell, J.P., Heller, M.J.: Rapid determination of single base mismatch in DNA hybrids by direct electrical field control. Proc. Natl. Acad. Sci. **94**, 1119–1123 (1997)
53. Marrazza, G., Chianella, I., Mascini, M.: Disposable DNA electrochemical sensor for hybridization detection. Biosens. Bioelectron. **14**, 43–51 (1999)
54. Drummond, T.G., Hill, M.G., Barton, J.K. Nature electrochemical DNA sensors. Biotechnology **21**, 1192–1199 (2003)
55. Bashir, R.: BioMEMS: state-of-the-art in detection, opportunities and prospects. Adv. Drug Deliv. Rev. **56**, 1–22 (2004)
56. Fritz, J., Cooper, E.B., Gaudet, S., Sorger, P.K., Manalis, S.R.: Electronic detection of DNA by its intrinsic molecular charge. Proc. Natl. Acad. Sci. **99**, 14142–14146 (2002)
57. Cui, Y., Wei, Q., Park, H., Lieber, C.M.: Nanowire nanosensors for highly sensitive and selective detection of biological and chemical species. Science **293**, 1289–1292 (2001)
58. Besteman, K., Lee, J.L., Wiertz, F.G.M., Heering, H.A., Dekker, C.: Enzyme-Coated carbon nanotubes as single-molecule biosensors. Nano Lett. **3**, 727–730 (2003)
59. Hintsche, R., Moller, B., Dransfeld, I., Wollenberger, U., Scheller, F., Hoffmann, B.: Chip biosensors on thin-film metal electrodes. Sensors Actuators B **B4**(3–4), 287–291 (1991)
60. Hintsche, R., Kruse, Ch., Uhlig, A., Paeschke, M., Lisec, T., Schnakenberg, U., Wager, B. Chemical microsensor systems for medical applications in catheters. Sensors Actuators B **B27**(1–3), 471–473 (1995)
61. Umek, R., Lin, M., Vielmetter, S.W., Chen, D.H.: Electronic detection of nucleic acids: a versatile platform for molecular diagnostics. Chem. Mol. Daign. **3**, 74–84 (2001)
62. Cia, X., Klauke, N., Glidle, A., Cobbold, P., Smith, G.L., Cooper J.M.: Ultra-low-volume, real-time measurements of lactate from the single heart cell using microsystems technology. Anal. Chem. **74**(4), 908–914 (2002)
63. Popovich, N.D., Thorp, H.H.: New strategies for electrochemical nucleic acid detection. Interface **11**(4), 30–34 (2002)
64. Zhu, H., Snyder, M.: Protein chip technology. Curr. Opin. Chem. Biol. **7**, 55–63 (2003)
65. Gangadharan, R., Anandan, V., Zhang, G.: Optimizing the functionalization process for nanopillar enhanced electrodes with GOx/PPY for glucose detection. Nanotechnology **19**, 395501 (2008)
66. Jia, F., Yu, C., Ai, Z., Zhang, L.: Fabrication of nanoporous gold film electrodes with ultrahigh surface area and electrochemical activity. Chem. Mater. **19**(15), 3648–3653 (2007)
67. Gooding, J.J., Praig, V.G., Hall, E.A.H.: Platinum-catalyzed enzyme electrodes immobilized on gold using self-assembled layers. Anal. Chem. **70**, 2396–2402 (1998)

68. Gooding, J.J., Erokhin, P., Losic, D., Yang, W., Policarpio, V., Liu, J., Ho, F., Situmorang, M., Hibbert, D.B., Shapter, J.G.: Parameters important in fabricating enzyme electrodes using self-assembled monolayers of alkanethiols. Anal. Sci. **17**, 3–9 (2001)
69. Losic, D., Shapter, J.G., Gooding, J.J.: Influence of surface topography on alkanethiol SAMs assembled from solution and by microcontact printing. Langmuir. **17**(11), 3307–3316 (2001)
70. Berchmans, S., Sathyajith, R., Yegnaraman, V.: Layer-by-layer assembly of 1,4-diaminoanthraquinone and glucose oxidase. Mater. Chem. Phys. **77**, 390–396 (2003)
71. Uang, Y.M., Chow, T.C.: Criteria for designing a polypyrrole glucose biosensor by galvanostatic electropolymerization. Electroanalysis **14**, 1564–1570 (2002)
72. Qiao, C.S., Tuzhi, P., Yunu, Z., Yang, C.F.: An electrochemical biosensor with cholesterol oxidase/sol-gel film on a nanoplatinum/carbon nanotube electrode. Electroanalysis. **17**, 857–861 (2005)

Chapter 2
Morphological Surface Modification

Abstract Surface modification approaches can be mainly regarded as by either a physical or a chemical means (note the biological ones are considered as a subset of chemical ones), or both. This chapter focuses on the physical means, particularly, a morphological approach to surface modification. Through illustration of how the area of a planar surface can be increased significantly by adding nanoscopic structures to the surface and that most of these nanostructures may be deemed useless for biosensor applications, it argues for a particular need of biosensors, namely, the need for aqua-robust nanostructures having the highest exposed surface area possible. Following that, it presents with full technical detail the processes to form aqua-robust nanopillar structures along with brief discussions on the underlying science. All the discussions are in the form of case studies, thus providing a practical guide for readers who might want to develop similar surfaces. This chapter ends with, again, a detailed discussion of a recently patented procedure on how to form integrated micro and nano structures on-a-chip for possible biosensor applications.

2.1 Increasing Surface Area with Nanopillars

With the advent of micro- and nanotechnologies, many flat (or planar) micro-electrodes have been used in the so-called micro total analysis systems (short for μ-TAS) [1]. Although improvement in detection limits is demonstrated with these μ-TAS systems, a reduction in current signal is also noted owing to the small size of the micro-electrodes. To overcome this problem, a common strategy is to increase the surface area of the electrode with nanostructures [2].

Nanostructures can provide increased surface areas due to their high surface-to-volume ratio. But when these nanostructures are formed on a planar substrate, the overall surface-area enhancement will be limited by the size of the underlying substrate. Then, how can one achieve a higher surface area when the size of the planar area (or the real estate) is fixed? The answer lies in a skyscraper metaphor, that is, to build up within a limited areal footprint [3]. Adding skyscraper nanostructures onto a planar surface offers a significant increase in its overall surface area when compared with the planar surface. This phenomenon can be illustrated by the example given in Fig. 2.1, where a 2D hexagonal array of vertically aligned

© Springer International Publishing Switzerland 2015

G. Zhang, *Nanoscale Surface Modification for Enhanced Biosensing*,

DOI 10.1007/978-3-319-17479-2_2

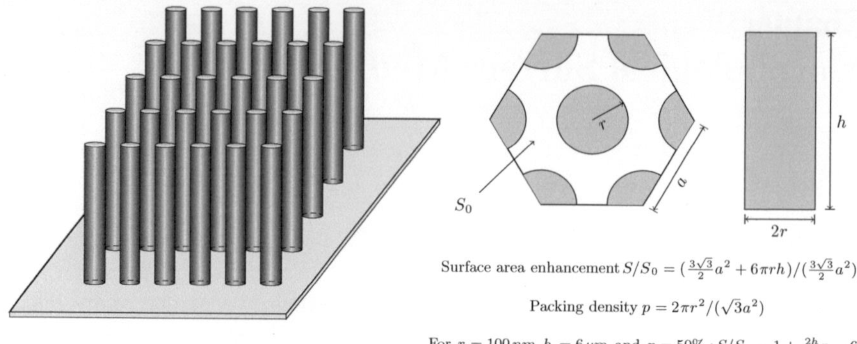

Surface area enhancement $S/S_0 = (\frac{3\sqrt{3}}{2}a^2 + 6\pi rh)/(\frac{3\sqrt{3}}{2}a^2)$

Packing density $p = 2\pi r^2/(\sqrt{3}a^2)$

For $r = 100\,nm$, $h = 6\,\mu m$ and $p = 50\%$: $S/S_0 = 1 + \frac{2h}{r}p = 61$

Fig. 2.1 Schematic illustration of surface area enhancement with the formation of nano pillar structures on a planar area

nanopillars is constructed on a planar surface [4]. The hexagon represents a unicell of the skyscraper nanopillars in a top view, and it contains a whole pillar in the center and 6 partial (1/3) pillars. Assuming that each pillar has a radius r and height h, and that the hexagonal unicell has a side length a, one can determine that the surface area of the hexagonal unicell is $S_0 = \frac{3\sqrt{3}}{2}a^2$ and that the enhanced surface area with the addition of nanopillars becomes $S = \frac{3\sqrt{3}}{2}a^2 + 3 \times 2\pi rh$, by accounting for the top and side areas of the nanopillars and the remaining base area in the unicell. With the introduction of packing density, which is defined as the ratio of the projection area of nanopillars and the hexagonal unicell area, $p = 3\pi r^2/(\frac{3\sqrt{3}}{2}a^2)$, we can write

$$S/S_0 = 1 + \frac{2h}{r}p \qquad (2.1)$$

This ratio is termed enhancement factor (it is the same as the roughness factor defined in Sect. 2.2.4) for the surface area. To get a quantitative sense for the surface area enhancement, let us consider a case where nanopillars have a radius of $r = 100$ nm, height of $h = 6\,\mu$m, and packing density of $p = 50\,\%$. By using Eq. (2.1) we can see that a 61-fold increase in the surface area can be achieved after the addition of these nanopillars.

To date, various skyscraper nanostructures have been fabricated using chemical vapor deposition (CVD) [5], physical vapor deposition (PVD) [6], and template based electrodeposition [7]. Lately, evidence has emerged to reveal that the nanotubes and nanorods developed by the CVD and PVD techniques could not sustain the capillary forces generated by nanostructure–liquid interaction [5, 6]. When vertically aligned nanostructures are exposed to a liquid environment, capillary forces will develop between the vertically aligned nanostructures and the liquid medium [8]. When the forces are large, the nanostructures will deform or bunch together. For example, the nanopillar structures fabricated by using PVD method

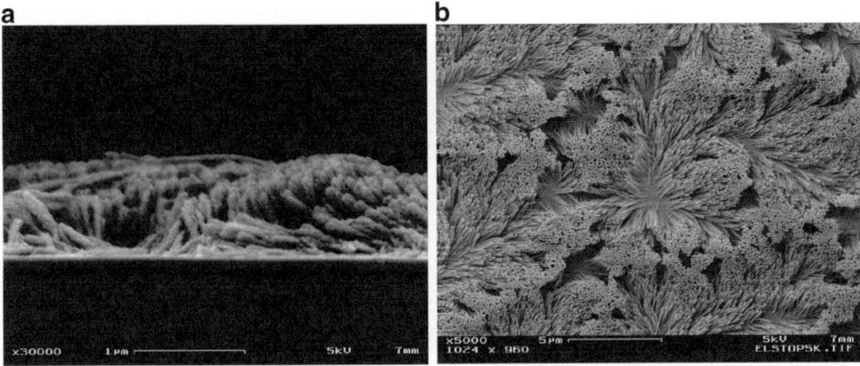

Fig. 2.2 SEM images of deformed silver nanopillars fabricated by PVD method: (**a**) a *side view* and (**b**) a *top view* of the deformed structure

in our lab deformed severely upon water exposure as shown in Fig. 2.2. Such a deformation in these skyscraper nanostructures will reduce the total surface area, thus posing a problem for their application in functional biosensor devices because a majority of biosensors will have to be exposed to aqueous environments. Therefore, to be useful as a component in a biosensor, these nanostructures need to have sufficient mechanical strength to overcome the capillary forces.

2.2 Fabricating Aqua-Robust Nanopillar Structures

To overcome this problem, robust skyscraper structures suitable for aqueous applications are necessary.

One easy and cost-effective solution is to use an aqueous based fabrication technique instead of a vapor based method to fabricate these structures. We have developed a template based electrodeposition technique to fabricate skyscraper nanostructures [9]. In this aqueous based fabrication method, porous anodic alumina (PAA) discs are used as templates to guide the electrodeposition of conducting materials through the pores of the PAA templates. In essence, this technique consists of two major steps: (1) making PAA templates and (2) fabricating nanopillar structures through electrodeposition.

2.2.1 Making PAA Templates

To fabricate porous anodized alumina (PAA) templates, one can use a two-electrode electrochemical cell to anodize aluminum sheets to form PAA. For example, we used the setup schematically depicted in Fig. 2.3 to do this, where a high purity

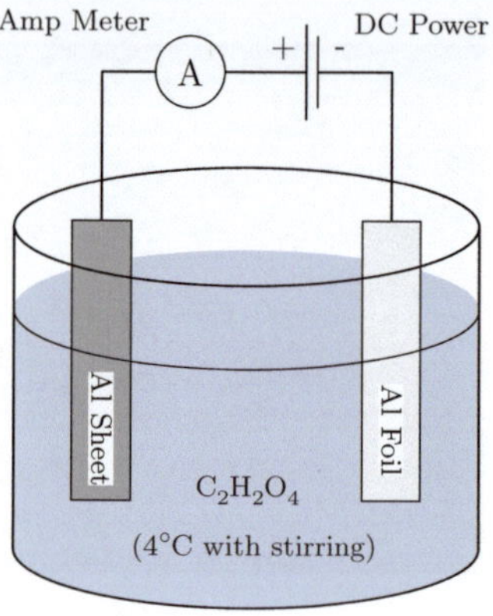

Fig. 2.3 Schematic of the setup for anodizing aluminum to form PAA template

(99.9 %) aluminum (Al) sheet (Alfa Aesar, MA) is used as the anode and a piece of commercial grade aluminum foil (Reynolds, VA) is used as the cathode [10]. For better PAA quality, some preparation steps are necessary. First, the aluminum sheet needs to be degreased in acetone and then cleaned by dipping it in 3.0 M NaOH solution till bubbling. Then, an electropolishing step is applied to the cleaned Al sheet in a solution of 10 % perchloric acid and 90 % ethanol by volume weight under an electrical potential of 20 V for about 1 min until mirror finishing is visible. After rinsed thoroughly in deionized water, the Al sheet is ready for anodization.

Typically, anodization of aluminum is achieved in 0.3 M oxalic acid ($C_2H_2O_4$) at 4 °C (typically, 3–5 °C) and 40 V in a two-step process. In the first step, the initial anodization is halted in about 1 h, and the Al sheet is disconnected from the circuit and the formed oxide layer during this time period is stripped by dipping the specimen in a solution of 6 %wt phosphoric acid and 1.8 %wt chromic acid at 60 °C for 30 min. This step will also improve pore formation and pore distribution in the subsequent anodization step. In the second anodization step, the Al sheet is reconnected into the circuit with power switched on for about 4 h. During anodization, it is necessary to rigorously stir the solution to enhance the mass transport to and from the Al sheet.

After anodization, the Al sheet becomes porous anodized alumina (PAA) containing parallel channels (or pores) running from the top surface of the sheet to the deep base of the sheet. These porous channels often have a diameter ranging from a few tens to a few hundreds of nanometer and are distributed in a hexagonal pattern. Figure 2.4 shows an SEM image of a typical PAA sheet where the zoom-in view highlights the orderly hexagonal distribution pattern of the formed pore channels.

For the anodized PAA to be used as templates, the non-conducting alumina layer at the bottom of each porous channel, commonly referred to as the barrier layer (see Fig. 2.5), needs to be removed such that all the porous channels are open at the bottom. This barrier layer can be dissolved by dipping the PAA sheet in 5 % H_3PO_4 solution at 30 °C for about 20 min.

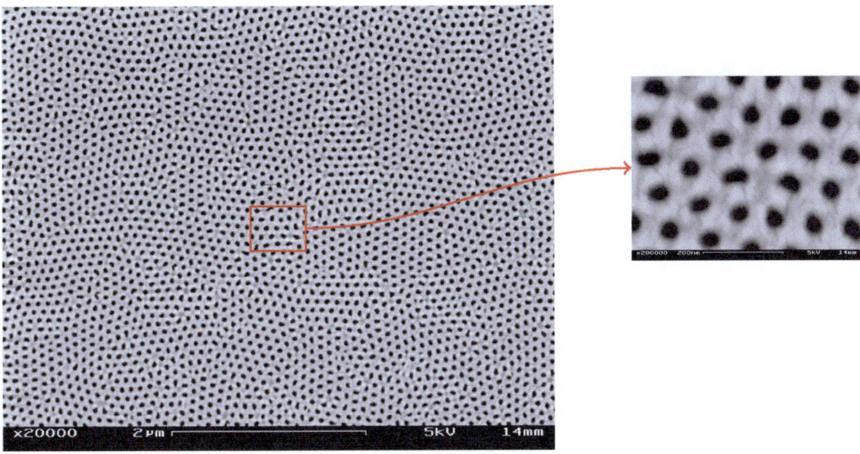

Fig. 2.4 SEM images of porous anodized alumina with a zoom-in view showing the hexagonal distribution pattern of the pores

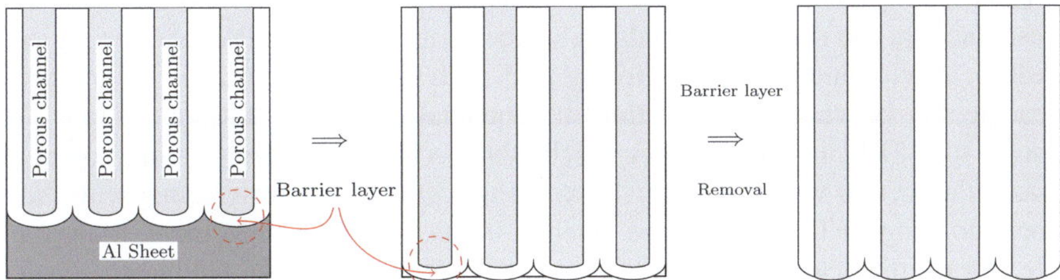

Fig. 2.5   Illustration of the formed barrier layer at the bottom of pore channels and its removal

2.2.2   Quantifying the Pore Dimensions and Their Relationships with Anodization Potential

The size of the formed pores can be determined directly from Scanning Electron Microscopy (SEM) images and the pore spacing (i.e., the center-to-center distance between two neighboring pores) can be quantified by using the fast Fourier transform (FFT) power spectra of the SEM images [10]. As illustrated in Fig. 2.6, in the case of a perfect hexagonal pore distribution (Fig. 2.6a), the FFT of such distribution is a 2D power spectrum in frequency domain consisting of six spectral spots (spots of high intensity) forming also a hexagon (Fig. 2.6b). Each spectral spot (two along one of the three orientations due to symmetry) of this hexagon represents the spatial periodicity (P) of the pore distribution in that orientation, and its distance to the center of the power spectrum determines the reciprocal of the spatial periodicity ($1/P$). Note that the spatial periodicity measures the same

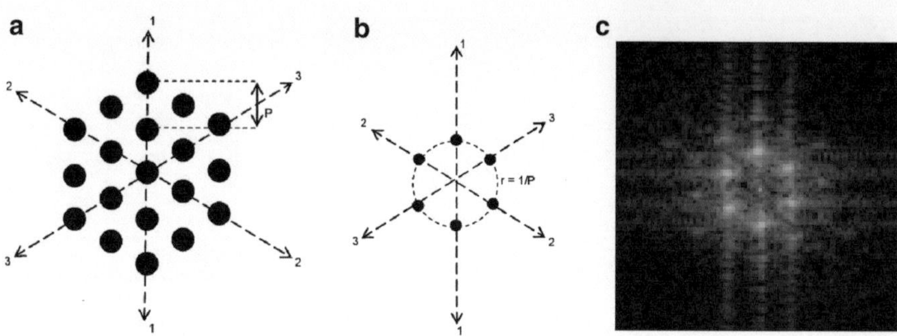

Fig. 2.6 (**a**) Illustration of a perfect hexagonal distribution of small dots or pores in spacial domain, (**b**) its 2D power spectrum after FFT in frequency domain, and (**c**) a resulting power spectrum after applying FFT to an SEM image of PAA

distance as the pore spacing. So when FFT is applied to SEM images (using Image J, National Institute of Health), the obtained 2D power spectra (see Fig. 2.6c) can provide quantitative information about the pore spacing in PAA.

To use the obtained PAA sheets as templates to fabricate 3D nanostructures it is desirable that the size and periodicity of the formed pores in PAA can be varied. Indeed, as we demonstrated previously [10], both the pore size (diameter) and pore spacing can be varied by adjusting the potential during anodization. Figure 2.7 shows four SEM images of PAA sheets obtained when the anodization potential was set at: (a) 20 V, (b) 30 V, (c) 40 V, and (d) 50 V, respectively, along with their corresponding FFT power spectra given in insets. From the SEM images, one can see that the diameter of the pores increases as anodization potential increases, and from the FFT spectra, it is clear that the diameter of the spectral rings decreases as the anodization potential increases, indicating that the pore spacing also increases as anodization potential increases.

Figure 2.7e shows some quantitative regression analyses for the relationships between the measured pore diameter and anodization potential and between pore spacing and anodization potential. From the regression results, we can see that both the pore diameter (PD) and pore spacing (PS) can be related to anodization potential (AP) approximately in a linear relationship as follows:

$$PD \simeq 1.35 \times AP \tag{2.2}$$

and

$$PS \simeq 2.55 \times AP \tag{2.3}$$

in which PD and PS are measured in (nm) and AP in (V). These linear relationships can serve as a convenient guide when deciding which anodization potential to use for a desired set of pore diameter and pore spacing.

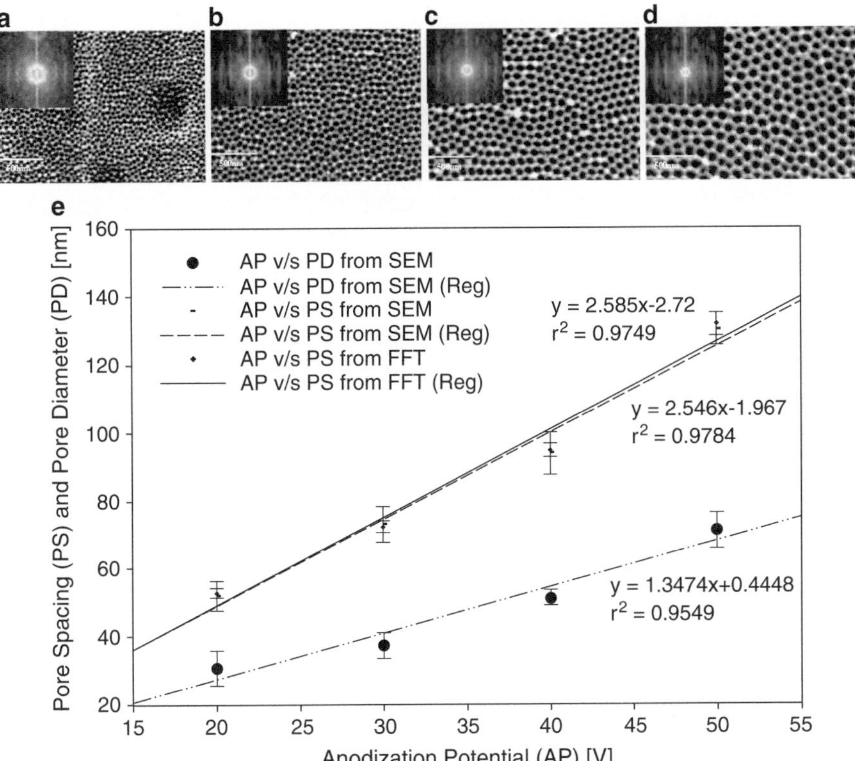

Fig. 2.7 PAA sheets obtained at various anodization potentials: (**a**) at 20 V, (**b**) at 30 V, (**c**) at 40 V, and (**d**) at 50 V. (**e**) Quantitative relationships between pore spacing or pore diameter and anodization potential along with statistical regression (Reg) curves

2.2.3   Fabricating Nanopillar Structures

By using the PAA sheets (after the barrier layer removal) as templates one can then use an aqueous electrochemical deposition (short for electrodeposition hereafter) method to fabricate nanopillar structures. Electrodeposition of gold nanopillars can be achieved by using a three-electrode electrochemical cell schematically shown in Fig. 2.8 driven by a potentiostat. In this three-electrode cell, a gold-coated PAA sheet is used as the working electrode (center), a platinum (Pt) wire gauze as counter electrode (right) and an Ag/AgCl electrode as the reference (left). A potentiostat is electronic device that provides controls of potential across the working electrode and the reference electrode and controls of current between the working electrode and the counter electrode.

In a typical fabrication process, a thin gold film (about 150 nm thick) is first sputter-coated onto one side of a PAA sheet to provide a conductive coating. Then a thicker gold layer (3 μm) is electrodeposited on top of the sputtered gold film to form a strong supporting base in Orotemp24 gold plating solution (Technic Inc, Cranston, RI) at a deposition current of $5\,mA/cm^2$ for about 2 min. The supporting base is then masked with Miccrostop solution (Pyramid plastics Inc., Hope, Arkansas) for insulation. After that, gold nanopillars are electrodeposited through the open pores of the PAA sheet from the uncoated side at a deposition current of $5\,mA/cm^2$ at 65 °C in the same plating solution. The deposition time can be varied for achieving nanopillars of different heights. After nanopillar deposition, the PAA template is dissolved in 2.0 M NaOH solution for 30 min. This procedure will result in a forest of gold nanopillars standing on a gold support base. Figure 2.8 also depicts three cross-section views of the specimen before plating, after plating, and after PAA template removal.

Figure 2.9 shows SEM images of some resulting nanopillar structures. Figure 2.9a shows a sample of silver nanopillars and Fig. 2.9b a sample of gold nanopillars. Since these structures are formed in an aqueous condition, the fact that the nanopillars in both cases exhibit no signs of collapsing or bunching (as we noted in Fig. 2.2) indicates the mechanical robustness of these nanopillars. In our earlier

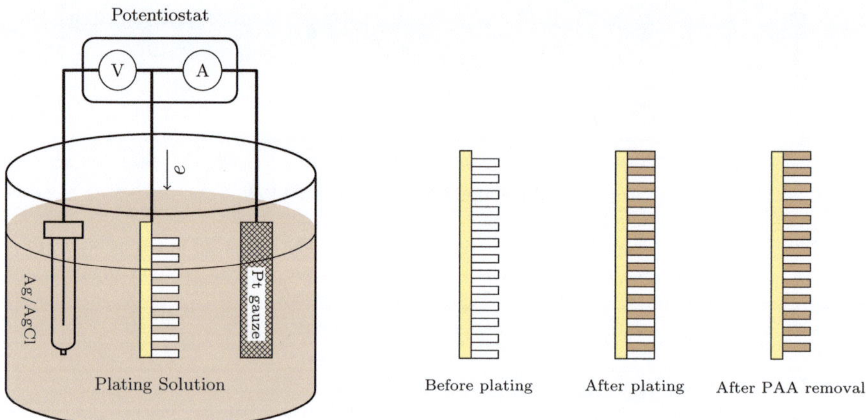

Fig. 2.8 Schematic of the setup and processing steps for electroplating the nanopillar structures

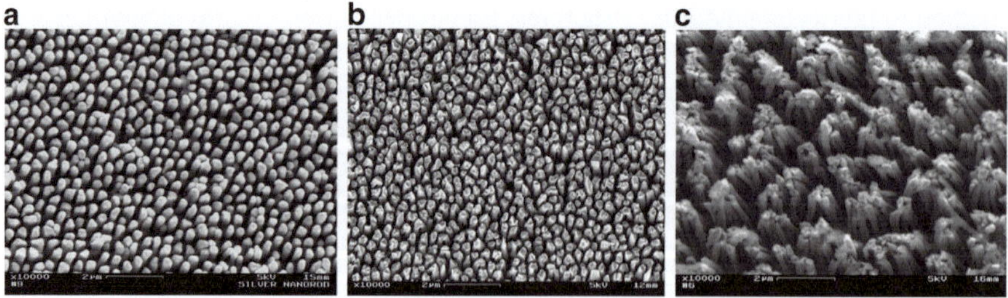

Fig. 2.9 SEM images of 3D nanopillar structures fabricated using an aqueous electrodeposition method along with PAA templates. (**a**) nanopillars with an aspect ratio = 10; (**b**) with an aspect ratio of 5; (**c**) with an aspect ratio of 30

study [9, 11] we noted that the mechanical strength of these nanopillars will weaken when the aspect ratio of these nanopillars becomes larger. As a reference point, the aspect ratio is 10 for the nanopillars in Fig. 2.9a and 5 for those in Fig. 2.9b. When the aspect ratio reaches 30 for gold nanopillars, the resulting structure exhibits a bunching phenomenon as one can clearly see in Fig. 2.9c. However, such a formation of the nanopillars remains unchanged after repeatedly subsequent aqueous exposures.

2.2.4   Confirmation for Surface Area Enhancement

As illustrated in Fig. 2.1, adding nanopillars to a planar surface can increase the overall surface significantly. To confirm that, we can perform electrochemical cyclic voltammetry (CV; see more discussion on the basics of CV in Sect. 3.3.2) to gold nanopillar structures in a solution of 0.3 M sulfuric acid (H_2SO_4). The idea here is to oxidize the exposed gold surface into gold oxide and then reduce it back to gold. Doing so allows us to quantify the amount of electrons required to reduce the exposed gold area by calculating the area under the reduction peak. Then by taking the ratio of the reduction peak of a nanopillar structure to that of a flat structure with the same project area, one can determine a surface roughness factor to represent the enhancement factor for the surface area. To do that, a sheet of nanopillar structure is used as the working electrode in a three-electrode electrochemical cell (see Fig. 2.10). For performing CV, a potentiostat is used to sweep the potential of the working electrode from -0.5 to $+1.5$ V again an Ag/AgCl reference electrode at a rate of 100 mV/s while the corresponding current with respect to the counter electrode is measured.

Fig. 2.10   A 3-electrode electrochemical cell commonly used for various electrochemical processes. The cell contains a working electrode, a reference electrode, and a counter electrode

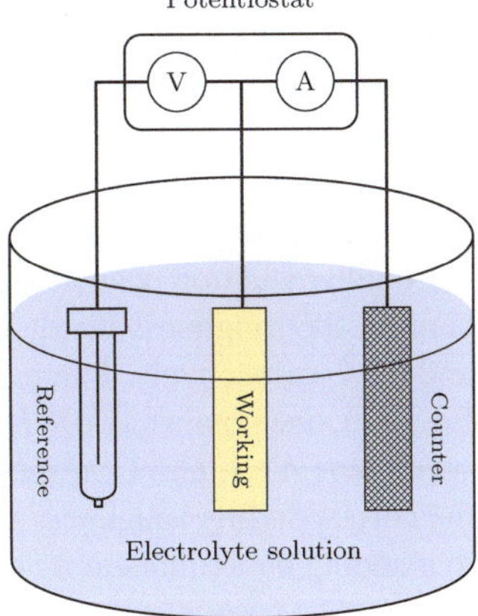

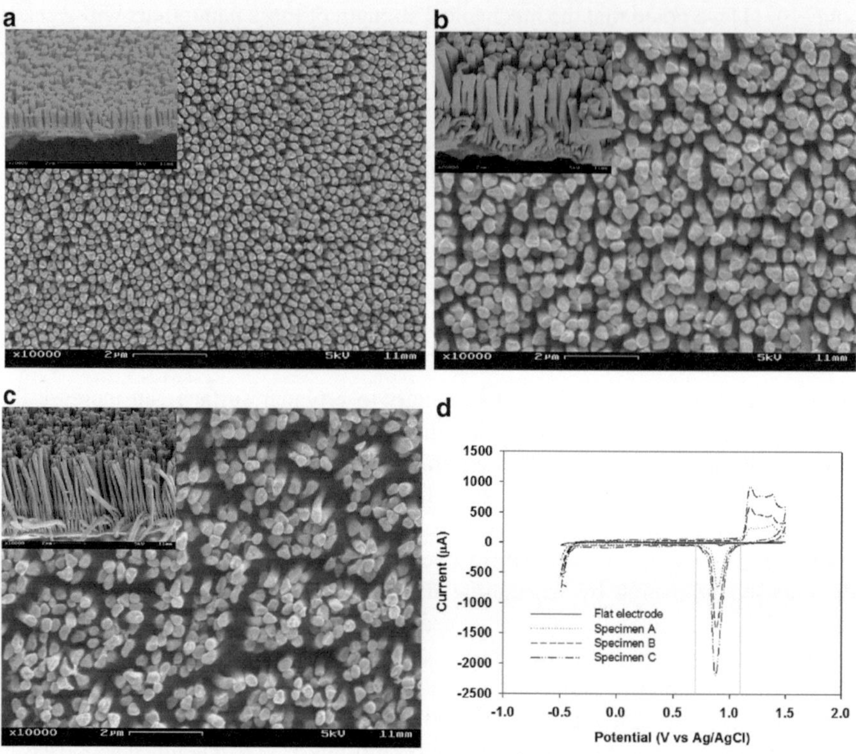

Fig. 2.11 SEM images of three nanostructure specimens, (**a**) Specimen A, (**b**) Specimen B and (**c**) Specimen C, along with insets showing a side-view of the specimens. (**d**) The corresponding cyclic voltammograms obtained for these specimens and a flat electrode as a reference

Figure 2.11 shows three samples of nanopillar structures with different nanopillar diameters and heights. Judging from the top and side views of the SEM images in each case, we estimated that $r = 60$ nm, $h = 1\,\mu$m, $p = 60\,\%$ for Specimen A, $r = 75$ nm, $h = 2.5\,\mu$m, $p = 55\,\%$ for Specimen B and $r = 75$ nm, $h = 6\,\mu$m, $p = 40\,\%$ for Specimen C. Figure 2.11d shows the obtained CV curves for these nanopillar structures along with a flat structure as a control. All these CV curves show an Au-oxide reduction peak at around 0.85 V, as expected. In comparison with the flat control, the roughness factor is found for specimens A, B, and C to be 20.0, 38.8, and 63.4, respectively. Interestingly, when plugging these dimensional values into Eq. (2.1), one obtains 21.0, 37.7, and 65.0 as the surface area enhancement factor for specimens A, B, and C, respectively. These enhancement factors are very close to the corresponding roughness factors determined from the CV, therefore confirming that adding nanopillars to a planar surface will indeed increase the surface area and that Eq. (2.1) provides a good estimate for the surface enhancement factor.

2.3 Fabricating Nanopillar Structures on-a-Chip

Although the surface area is drastically enhanced with the addition of these nanopillars, one challenge is the difficulty to integrate them into devices through further structural processing. For example, to use this type of nanopillar structures in biosensors, it is desirable that such structures could be directly formed on a chip, allowing further processes to be made to integrate these surface enhancing structures into microelectrodes. Although porous alumina thin films have been formed on glass or silicon substrates [12], their use as templates in forming standing nanostructures and further processing these structures into functional devices has rarely been attempted. To address this challenge, an on-chip based process has been developed [13, 14]. Here we highlight some key steps of this process.

2.3.1 Depositing Al/Au/Ti layers on a Glass Slide

To date, several different approaches have been used to produce PAA structures on silicon or glass substrates. All these approaches use a general procedure of depositing a thin film of aluminum on silicon or glass substrates with an underlying adhesion layer and sometimes an interfacial layer in between [15–19]. It is known that a very thin layer of Ti can assist the deposition of non-adherent metallic layers like Al but also withstand the electrochemical oxidation during the processes [20].

We noted in our work that a proper interfacial layer could play an important role in affecting not only the orderedness of the pores in PAA but also the removal of the formed barrier layer during anodization. As anodization progresses to reach the base of the Al layer, this barrier layer will reach the underlying conducting layer, separating the alumina layer from the conductive layer. If this non-conducting alumina barrier layer is not removed, it will obstruct electron transfer from the conducting base into the porous channels. Thus it must be removed before electrodeposition can be performed to fill the pore channels to form nanopillars.

To facilitate easy removal of the barrier layer, we found that adding a thin layer of gold on top of Ti layer is necessary, which, as a result, will also provide better electrical conductivity during electrochemical processes. The thickness of the gold layer is critical because a thicker layer would result in very high current toward the end of anodization, which could lead to severe pitting in the films or gradual dissolution of the alumina. In our experiments we found the thickness of this gold layer was better controlled about 20 nm.

As schematically illustrated in Fig. 2.12, to deposit metallic films on a chip, a glass slide is as the base chip in order to make the process more economical. To begin, a glass slide is first cleaned using RCA1 solution (H_2O:NH_4OH: $H_2O_2 = 5$:1:1) at 70 °C for 20 min. Then the glass slide is briefly etched in 0.5 %

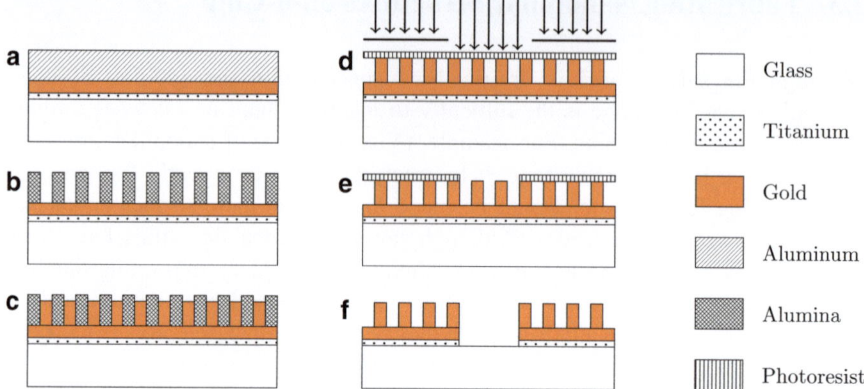

Fig. 2.12 Schematic of the processing steps for developing nanopillar structures on-a-chip. (**a**) Staring substrate of a glass slide coated with a titanium layer, a gold layer and an aluminum layer sequentially. (**b**) The top Al layer is anodized to form a nanoporous alumina template layer. (**c**) The nanoporous template is filled with gold via electrodeposition. (**d**) After the alumina template is removed, the gold structure is masked with a photo-resist layer and UV exposed to form the desired pattern via conventional photo-lithography. (**e**) The exposed photo-resist is developed. (**f**) The exposed gold structure is etched in solution

hydrogen fluoride (HF) solution for 30 s followed by rinsing in copious amount of water. As the first step (Step A in Fig. 2.12), a three-layer metallic film consisting of 5 nm Titanium (Ti), 20 nm gold (Au) and 800 nm Aluminum (Al) is deposited sequentially onto the glass slide using an E-beam evaporator (Kurt Lesker PVD 75, Kurt J. Lesker Company, Clairton, PA, USA). For these metallization steps, high purity (99.995 %) Al, Au Ti pellets are necessary, which can also be purchased from Kurt J. Lesker Company.

2.3.2 *Anodizing the Al Layer and Removing the Barrier Layer*

The three-layer metallic film formed on the glass slide is then masked using Microshield lacquer around the edges leaving only a central area of 25×25 (mm) exposed. The exposed area is further electropolished in a solution of $H_3PO_4{:}H_2SO_4{:}H_2O$ (2:2:1 wt) at 20 V. This step not only removes any oxides but also reduces the surface roughness. After that, an anodization step is applied to the top Al layer under a constant DC potential of 40 V in 0.3 M Oxalic acid at 4 °C for a few hours using the setup shown in Fig. 2.3.

The progress of anodization can be monitored by recording the current running to the Al layer. Typically, the current is at a constant level of around $2\,\text{mA/cm}^2$ throughout the anodization process. When the barrier layer reaches the gold layer the current will drop rapidly, indicating that all of the aluminum has been anodized.

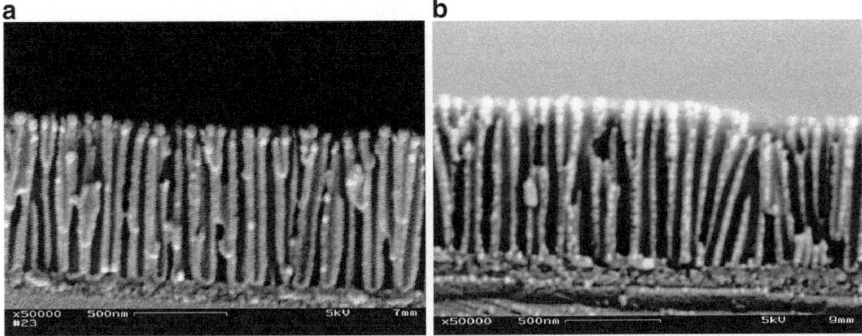

Fig. 2.13 SEM images of cross-section views of pore channels, (**a**) before barrier layer removal, (**b**) after barrier layer removal

At this stage the anodization reaches a critical moment. On the one hand, if the power (anodization potential) is cut off too soon, the barrier layer will remain intact, and on the other hand if the power is kept on for too long, the current will shoot up again due to breakage of the barrier layer at the bottom of all the pores. For example, we noted in our experiments that leaving the power on could cause the current to rise as high as 50 mA/cm^2 in a short period of time with the consequence of having all the pores dissolved completely.

A proper time to stop the anodizations, as we observed from experiments, is when the current rises from its lower level and reaches around 4 mA/cm^2. This ensures that the barrier layer at the bottom of all the porous channels can be removed easily by subsequent chemical dissolution. The end result of this step (Step B) is a layer of PAA siting above the Au and Ti layers. To remove the remaining barrier layer, the slide is placed in 5 % H_3PO_4 solution at 30 °C for 20 min. This step will dissolve the barrier layer, expose the underlying gold film at the bottom ends of all the porous channels, and simultaneously widen the pore slightly (e.g., from 40 nm after anodization to 50 nm after removal of the barrier layer). Figure 2.13 shows SEM images of a cross-section view of the porous channels before and after the barrier layer is removed.

2.3.3   *Electrodepositing Nanopillars and Removing PAA*

The formed PAA layer can now serve as a template for electrodeposition of nanopillars. The same process discussed in Sect. 2.2.3 can be applied here, except at a lower deposition current. To do that, the specimen is placed in an Au plating solution (Orotemp24, Technic Inc, Cranston, RI) under a constant current of 0.6 mA/cm^2 to fill the porous channels with gold to form Au nanopillars (Step C in Fig. 2.12).

The height of the nanopillars can be controlled by varying the deposition time. For example, we noted that 6-min deposition would yield nanopillars of approximately 600 nm in height. After electrodeposition, the PAA template is dissolved in 2 M NaOH solution.

2.3.4 A Film of Standing Nanopillars on-a-Chip and Its Further Processing into Micropatterns

The above three steps will result in a film consisting of a forest of vertically standing nanopillars on the Au/Ti layer coated on the glass slide (see Step D in Fig. 2.12). Such a chip structure can be used as electrode directly [21]. However, to further pattern it into desirable microelectrodes, conventional microfabrication procedures can be easily performed because this film of nanostructures is made of metallic materials of Au and Ti.

Prior to further patterning, the standing nanopillars on substrates need to be cleaned. This can be done by running electrochemical cyclic voltammetry (CV) in 0.1 M H_2SO_4 in a potential range from 500 to $+1{,}500$ mV until CV curves become stable using the 3-electrode cell shown in Fig. 2.10. The measured CV curves are often used to quantify the roughness factor of the nanopillar-modified surfaces. After that, these cleaned specimens are rinsed in deionized (DI) water.

To convert the nanopillars along with the underlying metallic film into patterned electrodes, a conventional photolithographic fabrication process can be applied. To do that (see Step D in Fig. 2.12), a positive photoresist (PR), Microposit 1818 (Rohm and Haas Electronic Material, Marlborough, MA, USA), is spun coated onto the surface and after a prebake step the desired pattern is transferred using in a mask aligner (MJB3, Karl Suss). After post-bake and PR development in developer MF-319, a solidified PR layer is formed masking the desired regions of nanopillars (Step E in Fig. 2.12). The unmasked nanopillars and the underlying metallic layers are etched. The final product is micro-patterned electrodes consisting of nanopillars standing on conducting Au film formed on top of a glass substrate (see Step F in Fig. 2.12). Figure 2.14 shows some SEM images of obtained structures, where a set of interdigitated microelectrodes (with width of about 5 μm) is formed. Three selective close views show that these microelectrodes are made from a film consisting of a forest of standing Au nanopillars on a glass slide chip. With further evaluation of the nanopillar film we estimate that $r = 25$ nm, $h = 600$ nm, and $p = 35.4\%$. Using Eq. (2.1) we find the surface enhancement factor of 18.

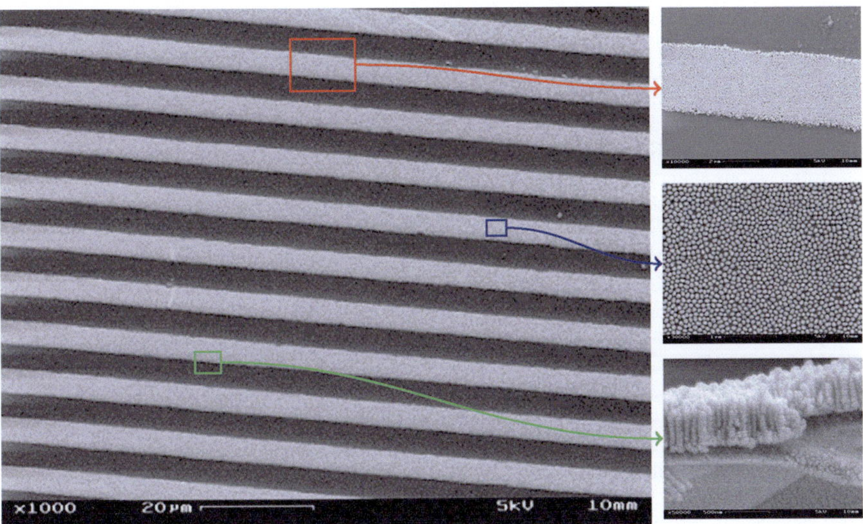

Fig. 2.14 SEM images of parallel electrodes etched on-a-chip, or a glass slide in this case. The three closer views highlight the fact that these microelectrodes are actually made of a film of a forest of vertically standing nanopillars

References

1. Xu, X., Zhang, S., Chen, H., Kong, J.: Integration of electrochemistry in micro-total analysis systems for biochemical assays: recent developments. Talanta **80**(1), 8–18 (1009)
2. Pumera, M., Escarpa, A.: Nanomaterials as electrochemical detectors in microfluidics and CE: fundamentals, designs, and applications. Electrophoresis **30**(19), 3315–3323 (2009)
3. Dunn, B., Long, J.W., Rolison, D.R.: Rethinking multifunction in three dimensions for miniaturizing electrical energy storage. Electrochem. Soc. Interface 17, 49 (2008)
4. Zhang, G.: Design and fabrication of 3D skyscraper nanostructures and their applciations in biosensors, in new perspectives in biosensors technology and applications. In-Tech. pp. 269–290 (2011)
5. Lau, K.K.S., Bico, J., Teo, K.B.K., Chhowalla, M., Amaratunga, G.A.J., Milne, W.I.: Superhydrophobic carbon nanotube forests. Nano Lett. **3**(12), 1701–1705 (2003)
6. Fan, J., Dyer, D., Zhang, G., Zhao, Y.: Nanocarpet effect: pattern formation during the wetting of vertically aligned nanorod arrays. Nano Lett. **4**(11), 2133–2138 (2004)
7. Xu, J., Huang, X., Xie, G.: Study on the structures and magnetic properties of Ni, Co-Al2O3 electrodeposited nanowire arrays. Mater. Res. Bull. **39**, 811–818 (2004)
8. Kralchevsky, P.A., Nagayama, K.: Capillary interactions between particles bound to interfaces, liquid films and biomembranes. Adv. Colloid Interf. Sci. **85**, 145–192 (2000)
9. Anandan, V., Yang, X., Kim, E., Rao, Y., Zhang, G.: Role of reaction kinetics and mass transport in glucose sensing with nanopillar array electrodes. J. Biol. Eng. **1**(1), 5 (2007)
10. Rao, Y., Anandan, V., Zhang, G.: FFT analysis of pore pattern in anodized alumina formed at various conditions. J. Nanosci. Nanotechnol. **5**(12), 2070–2075 (2005)
11. Anandan, V., Rao, Y., Zhang, G.: Nanopillar array structures for high performance electrochemical sensing. Int. J. Nanomed. **1**, 73–79 (2006)
12. Tanaka, Y., Sato, K., Shimizu, T., Yamato, M., Okano, T., Kitamori, T.: Biological cells on microchips: new technologies and applications. Biosens. Bioelectron. **23**(4), 449–458 (2007)

13. Zhang, G.: United State Patent US 8,453,319 B2
14. Zhang, G.: United State Patent US 8,865,402 B2
15. Zhao, G., Xu, C., Li, H.: Highly ordered cobalt-manganese oxide (CMO) nanowire array thin film on Ti/Si substrate as an electrode for electrochemical capacitor. J. Power Sources **163**(2), 1132–1136 (2007)
16. Balakrishman, S., Kripesh, V., Chong, S.: Fabrication of self-organized metal nanowire array using porous alumina template for off-chip interconnects. Int. J. Nanosci. **4**(4), 453–458 (2006)
17. Rabin, O., Herz, P.R., Lin, Y., Akinwande, A.I., Cronin, S.B., Dresselhaus, M.S.: Formation of thick porous anodic alumina films and nanowire arrays on silicon wafers and glass. Adv. Funct. Mater. **13**(8), 631–638 (2003)
18. Chu, S.Z., Wada, K., Inoue, S., Todoroki, S.: Fabrication and characteristics of nanostructures on glass by Al anodization and electrodeposition. Electrochim. Acta **48**(20–22), 3147–3153 (2003)
19. Sharma, G., Chong, S.C., Ebin, L., Hui, C., Gan, C.L., Kripesh, V.: Fabrication of patterned and non-patterned metallic nanowire arrays on silicon substrate. Thin Solid Films **515**(7–8), 3315–3322 (2007)
20. Hoogvliet, J.C., van Bennekom, W.P.: Gold thin-film electrodes: an EQCM study of the influence of chromium and titanium adhesion layers on the response. Electrochim. Acta **47**(4), 599–611 (2001)
21. Gangadharan, R., Anandan, V., Zhang, A., Drwiega, J.C., Zhang, G.: Enhancing the performance of a fluidic glucose biosensor with 3D electrodes. Sensors Actuators B **160**(1), 991–998 (2011)

Chapter 3
Biochemical Surface Modification

Abstract This chapter discusses the chemical side of surface modification with an emphasis on surface functionalization for electrodes used in biosensors. To continue the discussion of the last chapter, it presents two different approaches for the functionalization of surfaces modified morphologically with standing nanopillars. The first approach uses a conducting polymer to entrap enzymes to the surface through electrodeposition, and the second approach uses self-assembled monolayer alkanethiols to tether enzymes to the surface. In each approach, case studies are presented with full procedural details to showcase how the surfaces modified with nanopillars can be optimally functionalized through the tuning of relevant processing parameters. To make it easier to follow, this chapter embeds the basic knowledge of electrochemistry (e.g., amperometry, cyclic voltammetry, impedimetry, and enzymatic kinetics, etc.) throughout the text as information in-need or on-demand. From a sensing-element perspective, two types of molecules are discussed: catalytic enzyme molecules (e.g., glucose oxidase) and affinity-binding molecules (e.g., avidin-biotin couple). Through three example cases this chapter also discusses operations of biosensors in terms of target detection, signal measurement, data analysis, and quantification of detection sensitivity through calibration.

3.1 The Need for Surface Functionalization

In general, surface modification encompasses alterations of surface characteristics in a wide range of spectra including roughness, contact angle, surface charge, bonding strength, and reactivity, among others. For example, in Chap. 2 we discussed in detail the process of modifying surface morphology and roughness. The process to alter the surface property of a solid material to make it biologically active is sometimes referred to as surface functionalization. Surface functionalization in the context of biosensors usually means to attach a layer of biosensitive molecules such as enzymes onto the surface of an electrode.

Because these aqua-robust nanopillar structures discussed in Sect. 2.2 are fabricated using an electrodeposition process, the materials suitable for this process are limited to conductive materials such as metals and conducting polymers. These materials often lack biological activities. Thus for utility as electrodes in biosensors

© Springer International Publishing Switzerland 2015

G. Zhang, *Nanoscale Surface Modification for Enhanced Biosensing*,

DOI 10.1007/978-3-319-17479-2_3

these nanopillar structures need to be coated with a layer of biological sensitive molecules. Unfortunately, most of the biological molecules cannot be attached directly onto the nonactive surface of these nanopillar structures by a physical means, thus a proper chemical means is necessary.

3.2 Surface Functionalization Using Conducting Polymers

One straightforward approach is to further use an electrodeposition process with a conducting polymer. Conducting polymers such as polypyrrole and polyaniline have been extensively used for attaching (or immobilizing) enzymes like glucose oxidase to electrodes in glucose sensors. Using conducting polymers for functionalizing biosensor's electrodes is advantageous in that most of these polymers have good biocompatibility and minimum interference, and they can be easily electro-deposited on various metallic surfaces, thus making them good candidates for biosensor applications [1]. For example, polypyrrole has been used for functionalizing various electrode surfaces in amperometric glucose biosensors [2, 3]. The performance of this class of biosensors, however, is often affected by the various electropolymerization parameters used for the immobilization of glucose oxidase enzymes. So optimization of electropolymerization conditions of polypyrrole along with enzymes is necessary for achieving better biosensing results. Since most of the electrodes used in these glucose biosensors are not modified morphologically, our knowledge on the optimal conditions for electrodeposition of polypyrrole is limited to flat surfaces [4].

An electrochemical based glucose biosensor having electrodes with morphologically modified surfaces, especially with nanostructures, may have enhanced detection sensitivity, lowered detection limit, and even improved enzyme stability. For example, polypyrrole-coated glucose oxidase nanoparticles have shown increased sensitivity values [5] and carbon nanotube doped polypyrrole film showed better enzyme activity [6], albeit the disarrayed nanostructures used for surface modification in these cases may block the active sites of enzymes and impede the underlying mass transport. Surface modification using gold nanopillars, aligned carbon nanotubes and platinum nanowires fares much better by providing a microenvironment favorable for maintaining enzyme activity with increased enzyme exposure to the reactive species [7] and increased sensitivity [2, 8].

In this section we will discuss the importance of optimizing the functionalization process for a glucose biosensor using electropolymerization of pyrrole along with glucose oxidase when the electrode surfaces are morphologically modified with nanopillars. In particular, we will examine the effect of various electropolymerization parameters on the performance of the nanopillar-modified electrodes for glucose detection. The parameters considered include the roughness factor of the nanopillar-modified electrode surface, electrodeposition current density, and the total charge density during deposition, among others. The latter parameters determine the thickness of the polymer/enzyme film and the amount of enzyme,

Fig. 3.1 Electrochemical cell setup for electrode functionalization

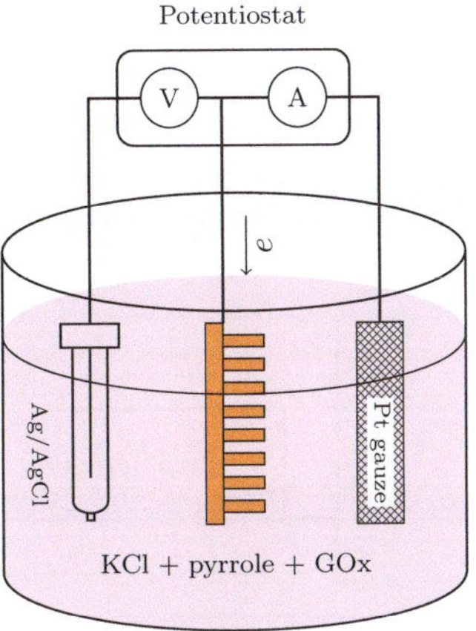

which in turn will affect the performance of the biosensor through alteration of electron transfer and mass transport behavior. The resulting glucose detection sensitivity is used as a metric for performance evaluation.

3.2.1 Experimental Procedure

For the experiments discussed here, pyrrole, β-D-glucose, glucose oxidase (GOx), and p-benzoquinone of analytical grade purchased from Sigma-Aldrich (St Louis, MO, USA) are used. All solutions are prepared using DI water. Figure 3.1 shows a simple setup for performing electrodeposition of pyrrole with enzyme (e.g., glucose oxidase, or GOx), consisting of a conventional three-electrode system with a Pt-gauze counter electrode, Ag/AgCl reference electrode, and a nanopillar-modified working electrode connected to a potentiostat. Prior to electrodeposition, the working electrode is cleaned in RCA1 solution (a mixture of H_2O:NH_4OH:H_2O_2 in 5:1:1 ratio) for organic residue removal. After that, electrodeposition of polypyrrole/GOx is performed under a galvanostatic condition in a 0.1 M KCl containing 0.05 M pyrrole and 0.5 mg/mL of GOx.

The electrodeposition parameters are optimized by varying the current density and time of deposition for achieving the highest possible glucose detection sensitivity for electrodes with nanostructure modified surfaces. Glucose detection is performed by running amperometry in a phosphate buffer solution (PBS) containing 3 mM p-benzoquinone as a mediator by measuring the amperometric current signals in response to incremental addition of glucose.

For glucose detection, a polypyrrole/GOx coated electrode is used as working electrode and it is held at a potential of 0.35 V versus Ag/AgCl reference electrode. Glucose solution is allowed to mutarotate for 24 h prior to experiments. Figure 3.2 illustrates the cascading chemical reactions in which glucose is oxidized by GOx

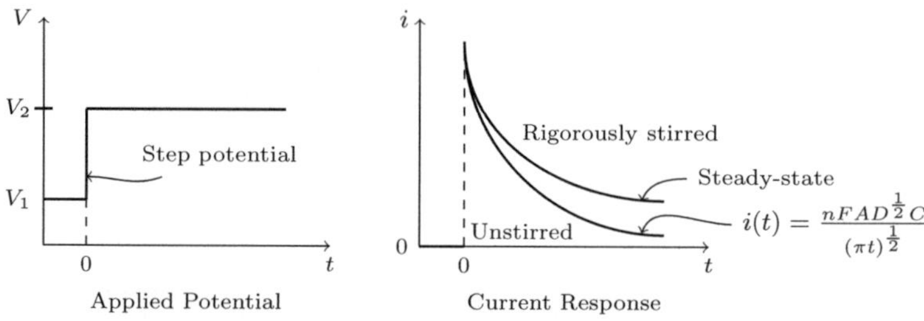

Fig. 3.2 Chemical reactions involved in glucose detection

Fig. 3.3 Illustration of chronoamperometry, often short for amperometry, in which a step potential is applied and the resulting current response in time is measured

mediated by the FAD/FADH$_2$ and the benzoquinone/hydroquinone couples for the generation of electrons. In each experimental run, after the background current stabilizes, a calculated amount of glucose from a 1 M glucose stock solution is added to the solution in an electrochemical cell for achieving an incremental increase in glucose concentration in the amount of 2.5 mM in the testing solution and the resulting amperometric current is measured after a fixed time period. To facilitate efficient mass transport for the enzymatic reaction, the solution is constantly stirred at a rate of 500 rpm. The detection sensitivity is determined from the Lineweaver–Burk plot through calibration of the relationship between the amperometric current and glucose concentration.

3.2.2 Some Basics on Amperometry

As illustrated in Fig. 3.3, amperometry is an electrochemical technique in which a step potential is applied to working electrode and the corresponding current response in time (hence, it is sometimes referred to as chronoamperometry) is measured. Because the applied potential often depletes an electroactive species at the surface of working electrode, it will cause a diffusion-driven mass transport of more electroactive species to the surface of working electrode. When the solution is unstirred (no convection), this diffusion-driven current response follows the Cottrell equation:

$$i(t) = \frac{nFAD^{\frac{1}{2}}C}{(\pi t)^{\frac{1}{2}}} \tag{3.1}$$

where n is the number of electrons exchange during electrochemical reactions, F is Faraday constant, A is surface area of the electrode, D is diffusion coefficient of the electroactive species, and C is concentration of the reactive specie in the bulk solution.

According to the Cottrell equation, when the solution is unstirred, the amperometric current eventually decays to zero. But at any given time, the current level is always linearly related to the concentration level of the analyte solution (C). Thus in performing amperometric experiments, if the solution is unstirred, it is better to record the current reading at a fixed time point. However, by introducing rigorous stirring to solution to accelerate the mass transport activity, namely adding a convection process to an otherwise diffusion-only process, one can obtain a steady-state current for the amperometric current response, which is still linearly related to C. So in the amperometric experiments discussed in this book, we keep the solution rigorously stirred.

3.2.3 Effect of Varying Surface Roughness Factor

The surface roughness factor, as defined in Sect. 2.2.4, which is the same as surface enhancement factor defined in Sect. 2.1 of an electrode having its surface morphologically modified with nanopillars can be mainly adjusted by the height of nanopillars. As illustrated in Sect. 2.1, an increase in nanopillar height will increase the surface area of the electrode, hence its roughness factor. This increase will in turn provide more surface area for facilitating electrochemical reactions, but will it affect the functionalization and sensing performance of the electrode? For this reason, we first examine the effect of surface roughness factor.

To minimize any compounding influence, the electrodeposition current density and the total charge density are kept the same while varying the surface roughness factor of the electrodes. For the current density, two different settings, 100 and $191\,\mu\text{A/cm}^2$, are used to ensure that the optimized condition is consistent when the current density is different. In both current conditions, the total charge density is fixed at $150\,\text{mC/cm}^2$ (meaning that the time of electrodeposition is different).

Figure 3.4 shows the variation of the measured amperometric current as a function of electrode's roughness factor in response to glucose under two deposition current density conditions (100 and $191\,\mu\text{A/cm}^2$) with the total charge density set at $150\,\text{mC/cm}^2$. The inset in the figure shows the cyclic voltammograms (CV) obtained for the corresponding electrodes. Clearly, all these CV curves exhibit an Au-oxide reduction peak in between 0.7 and 1.1 V. Since such a reduction peak is directly related to the actual electrochemically exposed surface area of each electrode, it can be used to quantify the actual surface area. Thus for each morphologically modified electrode, we calculate its roughness factor as the area under the reduction peak (by integrating the CV curve under the reduction peak) of a nanopillar-modified electrode to that of a flat electrode of the same projection area.

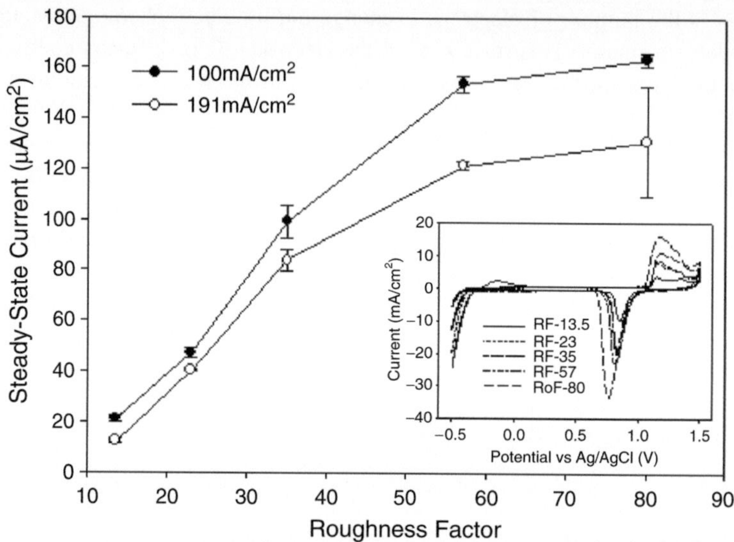

Fig. 3.4 Variation of the measured amperometric current with the roughness factor of surface modified electrodes ($N = 3$)

For the five electrodes tested here, their roughness factors are found to be about 13.5, 23, 35, 57, and 80, respectively, representing nanopillars of heights varying from about 1–8 μm.

From Fig. 3.4 it is seen that the measured amperometric current increases as the roughness factor increases and it appears to saturate when the roughness factor goes beyond 57. The same phenomenon is observed for both current cases. This current-saturation behavior at a higher roughness factor can be attributed to the difficulty encountered by glucose in diffusing down to the roots of the nanopillars for oxidation as the height of the nanopillars reaches a certain value [10]. It suggests that for glucose detection using the present method it is not necessarily beneficial to have electrodes modified with nanopillars that are too tall. Thus, to take full advantage of the nanopillar enhanced large surface area, we then select 60 as an optimal value for the roughness factor. Figure 3.4 also shows that the measured current is higher for the case under a lower deposition current density than for the case under a higher deposition current density, likely owing to a more uniform polypyrrol/GOx film formed under a lower deposition current density (see more discussion in the next section).

3.2.4 Effect of Varying Deposition Current Density

Figure 3.5 shows the variation of the measured current with deposition current density for electrodes with a roughness factor of about 60. When the total charge density is set at $150\,\mathrm{mC/cm^2}$ the measured current decreases as the deposition current density increases with no obvious peak performance within the range of

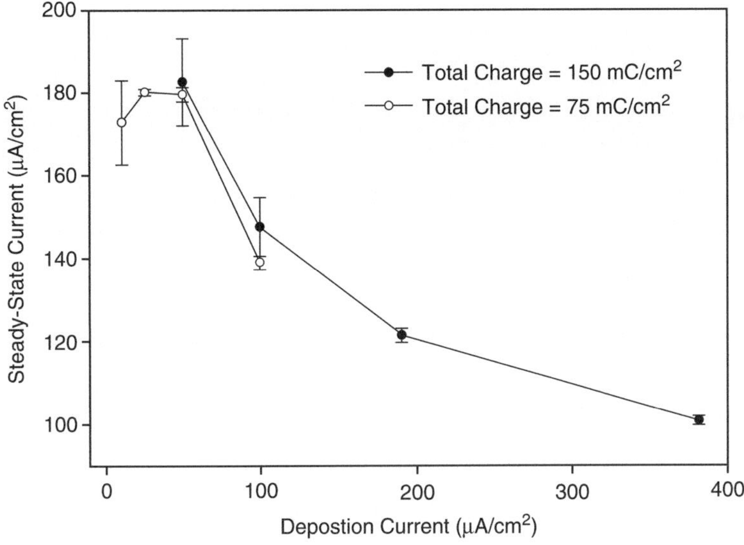

Fig. 3.5 Variation of measured amperometric current with deposition current obtained under two different total charges ($N = 3$)

the applied deposition current density (i.e., from 50 to $382\,\mu A/cm^2$). When the deposition current density is below $50\,\mu A/cm^2$ (achieved at a reduced total charge density of $75\,mC/cm^2$) it becomes obvious that the measured current does reach a peak value at the deposition current density of $50\,\mu A/cm^2$. From the overlap region it is seen that these two curves are not only very close to each other (the one under a lower charge density is slightly lower than the one under a higher charge density) but also following the same trend. Based on these results, we deem $50\,A/cm^2$ as an optimal value for the deposition current density, which is quite different from that obtained for the flat electrodes (i.e., $382\,\mu A/cm^2$) [4].

Such differences can be attributed to the presence of the nanopillar structures. Because of these closely spaced standing nanopillars, the mass transport of the electroactive species (including the pyrrole during electrodeposition and glucose during detection) to and from the electrode surface is expected to be different from that when nanopillars are absent. The evidence can be seen from the SEM images given in Fig. 3.6. Under a lower deposition current density of $50\,\mu A/cm^2$ all the nanopillars are covered by a thin and uniform layer of polypyrrol/GOx (see Fig. 3.6a; in fact, the uniform film is so thin that it looks as if no deposition occurred). When the deposition current density increases to $100\,\mu A/cm^2$, a slightly thicker film of polypyrrol/GOx is visible near the top ends of the nanopillars and a thinner film at the roots of the nanopillars (see Fig. 3.6b). This thick film formed during higher current densities surely will contribute to slowing down the diffusion of glucose, leading to lower current responses [9]. Under an even higher deposition current density of $382\,\mu A/cm^2$, the space between nanopillars close to the tips is all clogged-up by the polypyrrol/GOx deposition (see Fig. 3.6c). This presents a

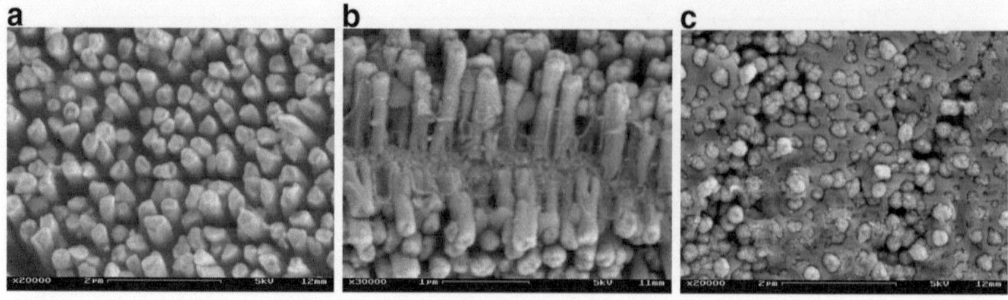

Fig. 3.6 SEM images of nanopillar-modified electrodes after electrodeposition of polypyrrol/GOx film when the total charge density is controlled at 150 mC/cm^2 under a deposition current density of (**a**) 50 μA/cm^2, (**b**) 100 μA/cm^2, and (**c**) 382 μA/cm^2

diffusion barrier for glucose to be detected. Based on these facts we believe that in the presence of closely packed standing nanopillars, a much lower deposition current density (e.g., 50 μA/cm^2 instead of 382 μA/cm^2) is necessary for uniformly electro-functionalizing the surface of these nanopillar-modified electrodes and for maintaining high efficiency in mass transport of glucose for its detection.

3.2.5   Effect of Varying Total Charge Passed for Deposition

By varying the charge density during electrodeposition, one can examine the influence of the amount of total charge consumed. For these experiments, all deposition is done under a fixed deposition current density of 50 μA/cm^2 to electrodes with a roughness factor of about 60. Figure 3.7 shows the obtained results. As charge density increases from 50 to 600 mC/cm^2, the measured current increases a little bit in the beginning and then decreases after reaching a peak value around 150 mC/cm^2. Thus, the value of 150 mC/cm^2 is regarded to be optimum for charge density for the electrodeposition.

Taken together, it is believed that an electro-process with a deposition current density of 50 μA/cm^2 and a charge density of 150 mC/cm^2 will provide an optimal condition for depositing polypyrrol/GOx film to functionalize the electrodes (with a roughness factor of about 60). These electro-processing parameters are quite different from those obtained for flat electrodes [4]. This fact suggests that chemical modification of a nanopillar-modified surface is quite different from that of a flat surface. Such differences can be attributed to the added surface area provided by the cylindrical walls of the nanopillars as well as the added difficulty for the active species in reaching the tiny space between these nanopillars [10].

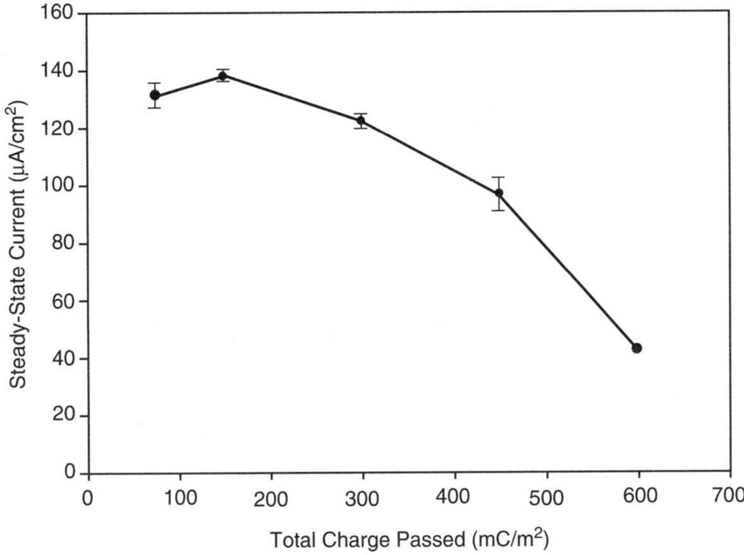

Fig. 3.7 Variation of measured amperometric current with the amount of charge in response to glucose ($N = 3$)

3.2.6 *Calibration for Detection Sensitivity*

To complete the discussion for this section, let us take look how the detection sensitivity is obtained through calibration of the relationship between measured amperometric current and analyte concentration. Figure 3.8 shows some typical measurements in time of the amperometric current response when the concentration of glucose is increased in increments of 2.5 mM successively. Note the stepwise steady-state amperometric current levels in response to each drop of glucose addition.

By taking the current at each glucose concentration level, a scattered-plot is made and used to calibrate the relationship between the measured current and the glucose concentration. As shown in the inset in Fig. 3.8, the calibration curve exhibits a very good linear relationship ($R^2 = 0.9998$) between the measured current and glucose concentration in a range from 2.5 to 15 mM (note that a nominal range for physiological glucose concentration is between 3.5 and 6.5 mM [11]). Based on the slope of this calibration curve, a sensitivity value is then determined by dividing the slope value with the projected area of the electrode. For this case, the sensitivity is found to be $36\,\mu A \cdot mM^{-1} \cdot cm^{-2}$.

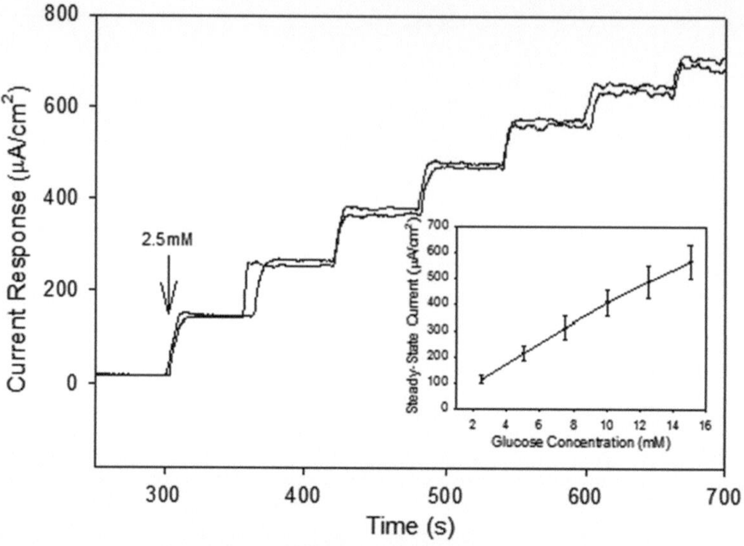

Fig. 3.8 Amperometric current responses to successive additions of glucose for a nanopillar-modified electrode. *Inset* shows the current-concentration calibration curve

3.3   Surface Functionalization Using Self-Assembled Monolayers

An alternative to using conducting polymers is to use alkanethiol self-assembled monolayers (SAM) as anchoring molecules to tether sensitive molecules to electrodes for their functionalization. Immobilization of enzymes onto electrode surfaces using SAM of alkanethiols has been commonly used because SAM molecules offer easy formation of well-ordered and stable monolayers of molecules for anchoring a wide range of biosensitive molecules [12, 13]. This is one advantage that SAM has over conducting polymers because the number of biosensitive molecules that are compatible with the electrodeposition process is limited.

With SAM, a similar optimization question remains and it is about what type of alkanethiol to use since they come with different chain lengths. As reported in the literature [14, 15], a longer chain length (e.g., 11-mercaptoundecanoic acid, or MUA) produces a more ordered assembly of molecules on a flat-surface electrode with a higher degree of surface coverage and less defects than a shorter chain length (e.g., 3-mercaptopropionic acid, or MPA). But MPA SAM on a flat electrode exhibits a lower electron transfer resistance than MUA SAM and gives rise to higher detection sensitivity than MUA SAM. Since the surface coverage of these SAM layers mainly depends on the surface morphology of the electrodes [16], it is believed that the presence of the closely packed standing nanopillars in the surface modified electrodes may alter the formation of the alkanethiol SAM at the surface. Thus it is logical to ask: which type of alkanethiol SAM, a short chain or a long chain, will help facilitate a better sensing performance for nanopillar-modified electrodes. To answer this question, we will examine two alkanethiol SAMs (i.e., the MPA and MUA) as anchoring molecules for the functionalization of nanopillar-modified electrodes. Note that MPA [3-mercaptopropionic acid: $HS-(CH_2)_2-COOH$] and MUA [11-mercaptoundecanoic acid: $HS-(CH_2)_{10}-COOH$] differ only in their chain length by eight CH_2 groups.

3.3.1 Experimental Procedure

Prior to SAM formation, the surfaces of nanopillar-modified electrodes need to be cleaned. It is usually done by performing cleaning using RCA1 solution ($H_2O:NH_4OH:H_2O_2 = 5:1:1$) to remove any organic residues. After that, electrochemical cleaning is performed by running cyclic voltammetry (CV) within a potential range from -0.5 to $1.5\,V$ in $0.3\,M$ H_2SO_4 solution at a scan rate of 100 mV/s until a reproducible CV curve is obtained. After that, the electrodes are placed in ethanol solution containing $10\,mM$ of either the MPA or MUA molecules for 24 h followed by washing in ethanol solution. SAM formation on these electrodes can be characterized via electrochemical means including the CV and electrochemical impedance spectroscopy (EIS) techniques at $25\,°C$ using a three-electrode electrochemical cell (like the one shown in Fig. 3.1 with different solutions).

For characterizing the surfaces functionalized with SAMs, CV measurements are taken by scanning the potential from -0.2 to $0.6\,V$ at a scan rate of $100\,mV/s$ and EIS is performed in a frequency range from $0.1\,Hz$ to $100\,KHz$ with an alternating-current signal of $10\,mV$ amplitude over the formal potential of the redox couple used. The tests for characterizing SAM formation are often done in solution of $0.1\,M$ phosphate buffered solution (PBS, pH 7) containing $2\,mM$ $Fe(CN)_6^{3-/4-}$ (ferri:ferro $= 1:1$) mixture as the redox couple.

To quantify the percent defect in the SAM coverage, the same CV tests but in different solution (i.e., $0.1\,M$ H_2SO_4) are performed by scanning the potential from -0.5 to $1.5\,V$ at a scan rate of $100\,mV/s$. The obtained voltammetric reduction peaks associated with the uncovered areas (i.e., the exposed gold oxide) of the SAM treated electrode surfaces are evaluated. The ratio of the uncovered area of a SAM treated electrode to that of a bare control electrode is calculated as the percent defect in the SAM layer.

To quantify the surface coverage (Γ) of the SAM molecules, stripping CV measurements [14, 17, 18] are obtained in 0.1 M NaOH within a voltage range from -1.6 to $-0.2\,V$ at a scan rate of $100\,mV/s$. In these tests, the voltammetric reduction peaks associated with SAM desorption are evaluated. To do that, CV experiments are performed and from a reduction peak, the amount of charge is calculated by first integrating the area under the reduction peak and then offsetting the value by that of a bare nanopillar-modified electrode. With the formula $\Gamma = Q/nFA$ [17], the surface coverage of SAM molecules can be determined. In the formula Q is the amount of charge, $n(=1)$ is the number of electrons involved in the reaction, F ($=96{,}485$ C/mol) is the Faraday constant, and A ($=0.04$ cm^2) is the electroactive surface area.

The SAM layer here only serves to provide anchors for enzyme immobilization for the functionalization of the surfaces of these nanopillar-modified electrodes. To functionalize them, the carboxyl groups in the SAM need to be activated in a freshly prepared solution of 0.1 M 2-(N-morpholino)ethanesulfonic acid containing 75 mM 1-ethyl-3-(3-dimethylaminopropyl) carbodiimide and 15 mM N-hydroxysuccinimide buffered at pH 4.5 for 2 h. This step is to turn the COOH groups of the SAM molecules into reactive N-hydroxysuccinimide esters. After washing in 0.1 M PBS the activated electrodes are placed in 0.1 M PBS containing 1 mg/mL of the GOx with constant stirring for another 2 h for GOx immobilization. For better performance, the functionalized electrodes are washed thoroughly with 0.1 M PBS and stored at 4 °C in 0.1 M PBS solution at pH 7.0 prior to testing.

For evaluating the sensing performance of these SAM treated and GOx functionalized nanopillar-modified electrodes, their amperometric currents in response to glucose at various concentrations are quantified. These experiments are performed in 20 mL 0.1 M PBS containing 3 mM p-benzoquinone with addition of various amount of 1 M glucose using a three-electrode electrochemical cell (see reactions in Fig. 3.2). During these tests, the electrodes are held at a constant potential of 0.35 V. In each test, the background current is allowed to stabilize before a drop (50 μL) of 1 M glucose is added to the solution, and after the amperometric current response stabilizes, another drop of glucose is added and the corresponding current response is measured until a new steady state is reached. In this manner, each incremental addition of glucose to the solution results in an equivalent increase in glucose concentration by 2.5 mM.

3.3.2 Some Basics on CV and EIS Experiments

As illustrated in Fig. 3.9, cyclic voltammetry is a technique in which the potential is cycled from an initial value (V_1) to a predetermined value (V_2) and then back to the initial (in a triangle-shaped profile). The resulting current-potential response often reveals some important information about the kinetics of the oxidation and reduction reactions, including the magnitude of the peak current (i_p) and the anodic (or oxidation; $E_{p,a}$) and cathodic (or reduction; $E_{p,c}$) peak potentials. According to the Randles–Sevcik equation, the peak current can be approximated as

$$i_p \approx 269 n^{\frac{3}{2}} A D^{\frac{1}{2}} C v^{\frac{1}{2}} \ \ (\text{mA}) \tag{3.2}$$

where v is the scan rate in V/s and the rest of the parameters are the same as in Eq. (3.2). The values for the peak potentials are often governed by the nature of the

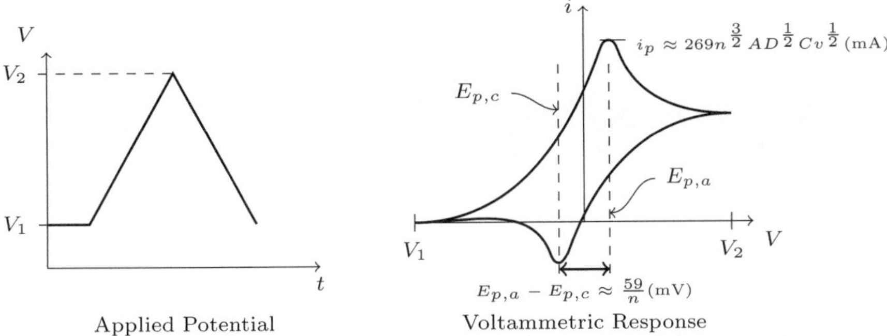

Fig. 3.9 Illustration of cyclic voltammetry, often short for CV, in which the potential is cycled from V_1 to V_2 and back in a triangle-shaped profile and the resulting current versus potential response is measured

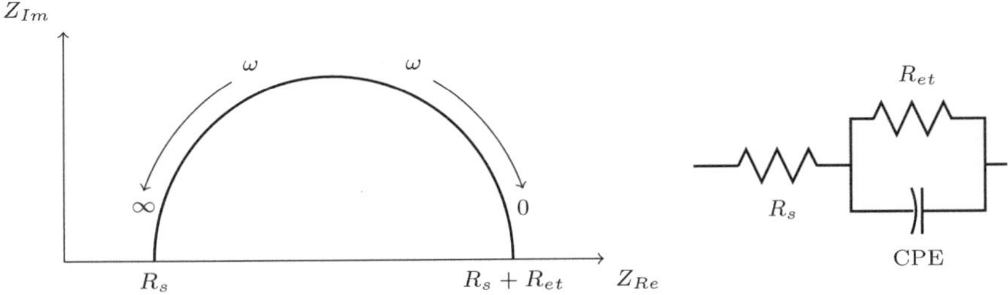

Fig. 3.10 Illustration of a Nyquist plot representing the complex impedance measured from electrochemical impedance spectroscopy, often short for EIS. A Randles equivalent circuit is often used to interpret the Nyquist plot

reactions and the state of mass transport. For a totally reversibly reaction, one often finds the following relationship:

$$\Delta E_p = E_{p,a} - E_{p,c} = \frac{59}{n} \ (\text{mV}) \tag{3.3}$$

Figure 3.10 shows a typical Z_{Re}–Z_{Im} plot obtained from an EIS experiment in which a sinusoidal perturbation potential with varying frequency is applied to working electrode and the resulting complex impedance ($Z_{Re} + iZ_{Im}$ or $Z' + iZ''$) is determined. The obtained complex impedance is often presented in a Nyquist plot in a Cartesian coordinate in which Z_{Re} or Z' is the abscissa and Z_{Im} or Z'' is the ordinate. To interpret the Nyquist plot, a Randles equivalent circuit, shown also in Fig. 3.10, is often used. This Randles circuit consists of a resistor (representing the solution resistance, R_s) in series with another resistor (representing the electron-transfer resistance, R_{et}) which is connected in parallel with a capacitor (representing constant phase element of the solution and electrical double layer, short for CPE). By fitting the Randles circuit model to the Nyquist plot, which in many situations

Fig. 3.11 SEM image of a typical nanopillar-modified electrode after SAM treatment and Gox functionalization

is a semicircle, one can determine the values for R_s, R_{et}, and CPE. In referring to the Nyquist semicircle shown in Fig. 3.10, one can estimate the values of R_s and R_{et} quickly. For example, the left intersect of the semicircle with the abscissa is R_s and the diameter of the semicircle is R_{et}.

3.3.3 Characterization of SAM Formation

Figure 3.11 shows an SEM image of a typical nanopillar-modified electrode after SAM treatment. From this image, the diameter of the nanopillars is estimated to be around 200 nm and the height around 2.5 μm. The roughness factor for these electrodes is calculated to be approximately 45 from electrochemical cleaning CV curves.

Figure 3.12 shows the CV curves obtained for a bare, and MPA and MUA treated electrodes evaluated in the presence of the redox couple. Clearly, both the bare and MPA treated electrodes exhibit peak-shaped CV curves with a peak-to-peak separation (ΔE_p) of approximately 59.8 mV for the oxidation and reduction of $Fe(CN)_6^{3-/4-}$. This value is very close to the ΔE_p of an ideal Nernstian one-electron reversible reaction of 59 mV (see Eq. (3.3)), indicating a highly efficient electron-transfer mechanism across the electrode/electrolyte interface of a nanopillar-modified electrode. However the current level for the MPA treated case is almost a half of that for the bare case, suggesting that while the electron

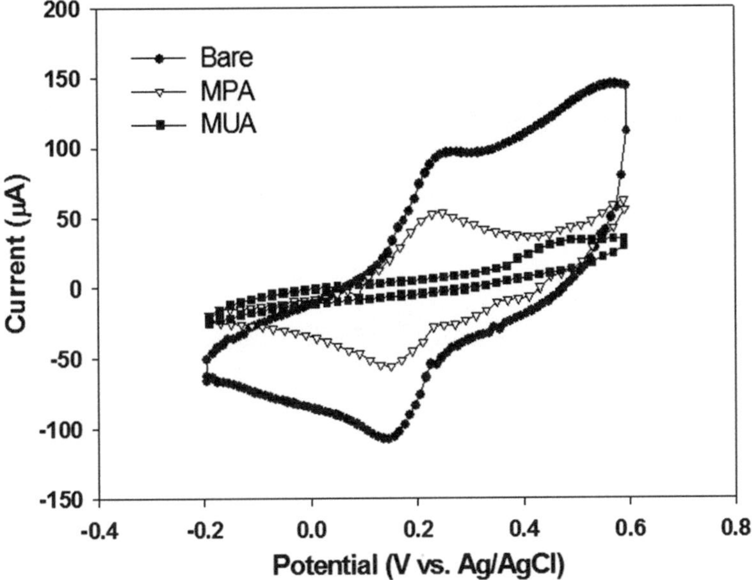

Fig. 3.12 CV curves obtained for a bare, MPA and MUA treated electrodes evaluated with $Fe(CN)_6^{3-/4-}$ as the redox couple

transfer efficiency is not affected the area available for facilitating electron transfer is significantly reduced. Moreover, these double-peak shaped CV curves are indicative of a diffusion-controlled electrode process.

For the MUA treated case, the shape of its CV curve is quite different: it looks much like a sigmoid curve having significant hysteresis between the forward and backward scans. This behavior implies that the adsorption of MUA molecules has significantly lowered the electron transfer rate such that the electrode process is no longer controlled by a diffusion process but by a kinetics-controlled process, meaning that it is limited by the rate of electron transfer. These results indicate that both the MUA and MPA molecules form SAM structures covering the electrode surface and that there are more MUA molecules than MPA molecules blocking the pathways for electron transfer across the electrode–electrolyte interface for facilitating redox activities, owing possibly to the longer chain length of the MUA molecules forming more lateral molecular bonds.

Figure 3.13 shows the corresponding impedance spectra (Nyquist plots) for these electrodes. The two SAM treated electrodes show semicircular Nyquist plots whereas the bare electrode exhibits an almost straight line plot (see the inset plot in Fig. 3.13). Since a semicircular feature is indicative of the presence of solution resistance and electron transfer resistance (see Fig. 3.10), these results point to the existence of blockage for electron transfer across the electrode/electrolyte interface, confirming the formation of SAM molecules on the electrode surfaces. Moreover, the MUA treated electrode exhibits a larger semicircle than the MPA treated

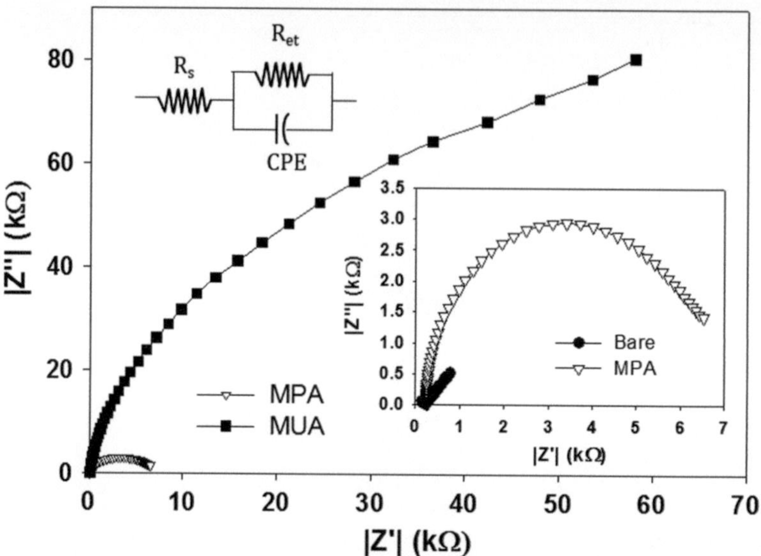

Fig. 3.13 Nyquist plots from the impedance measurements for the same electrodes with a close-up view of the low impedance range given in the *lower inset* and a Randles equivalent circuit in the *upper inset*

Table 3.1 Resolved R_s and R_{et} values from the Randles circuit (fitting errors given in parentheses)		

Electrodes	R_s (Ω)	R_{et} (Ω)
Bare	227.0 (2.0 %)	589.5 (5.0 %)
MPA treated	256.6 (0.9 %)	6281.0 (1.7 %)
MUA treated	229.0 (1.0 %)	209370.0 (4.3 %)

electrode, suggesting a high degree of SAM coverage for the MUA than for the MPA molecules. To quantify the solution resistance and electron transfer resistance, the Randles equivalent circuit, shown again as an inset in Fig. 3.13, is used to fit the obtained semicircular Nyquist plots to determine the values for R_s and R_{et}. As listed in Table 3.1, R_{et} obtained for the MUA treated electrode is much higher (about 27 times) than that for the MPA treated electrode, while R_s changes only slightly. This fact confirms that the MUA molecules indeed post a higher electron transfer resistance at the electrode surface than the MPA molecules.

Figure 3.14 shows the CV curves obtained for evaluating the gold-oxide reduction peak for the SAM treated electrodes along with a bare control in H_2SO_4. All these CV curves exhibit an Au-oxide reduction peak at around 0.78 V, indicating that all these electrodes possess a certain amount of exposed gold oxide, or defects in SAM coverage. By the ratio of the area under the reduction peak (through integration of the CV curve under the peak) of a SAM treated electrode to that of the bare electrode, a measure of the percentage of defect is determined. As listed in Table 3.2, the percent defect is approximately 87.3 and 37.8 % for the MPA and MUA cases, respectively. These values are high when compared with flat electrodes: 52 % for the MPA and 0 % for the MUA [15].

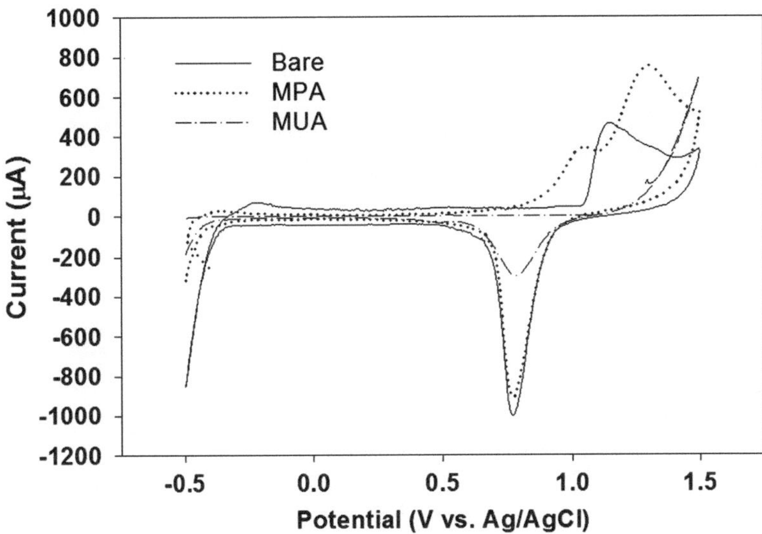

Fig. 3.14 CV curves obtained for bare, MPA and MUA treated electrodes in quantifying the percent defect in SAM coverage in 0.1 M H_2SO_4

Table 3.2 Evaluation of surface coverage and percent defect

SAM	Γ (10^{-8} mol/cm^2)	% Defect	% Adsorption
MPA	1.38 ± 0.1	87.3	12.7
MUA	2.37 ± 0.3	37.8	62.2

Figure 3.15 shows the CV curves obtained for evaluating the voltammetric reduction peak associated with desorption of MPA and MUA molecules. From these CV curves, two peak currents are visible for both the MPA and MUA treated electrodes. We believe that the peaks at around -0.82 V for MPA and at around -1.03 V for MUA are due to the cleavage of gold-sulfur bonds. This observation is consistent with the reported results in literature. For example, a peak desorption current between -0.6 and -0.9 V is found for short alkanethiols ($n = 2$–6) and between -1.0 and -1.2 V for long alkanethiols ($n = 11$–18) [14, 18–20]. The nature of the second peak at a higher reduction potential in both curves is unclear although it was also observed by others with flat surfaces [20]. Based on these desorption peak currents, the desorption charge was determined by integrating the reduction peak from -0.8 to -0.9 V for the MPA treated electrode and from -1.0 to -1.2 V for the MUA treated electrode. The surface coverage, Γ, is then calculated for MUA and MPA treated electrodes as listed in Table 3.2. Comparing with the reported values for the surface coverage of MPA (5.12×10^{-10} mol/cm^2) and MUA (8.30×10^{-10} mol/cm^2) on flat surfaces [15], the values for nanopillar-modified electrodes are roughly 27 and 28 times higher, respectively. This increase can be attributed to the increase in the electroactive surface area in the nanopillar-modified electrodes. This increase, however, does not correspond to the actual increase in the surface area of 45 times. This phenomenon may be attributed to the high percentage

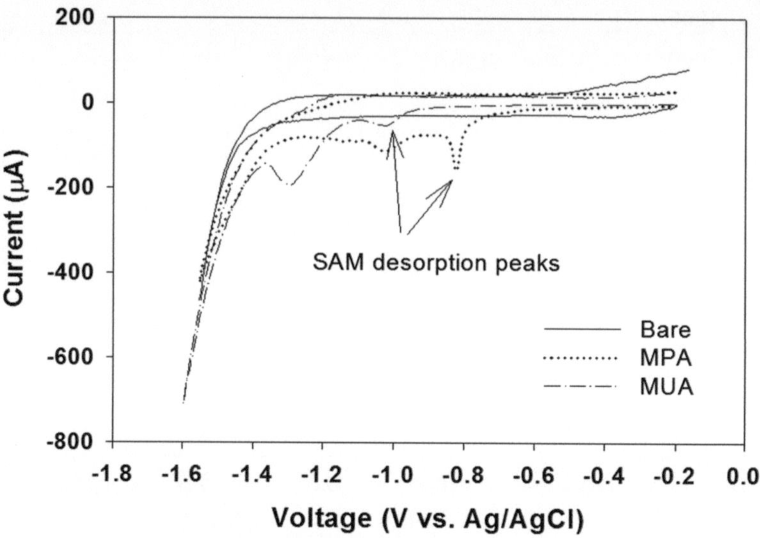

Fig. 3.15 CV curves obtained for bare, MPA and MUA treated electrodes in evaluating SAM desorption in 0.1 M NaOH

of defects in the SAM structures on the nanopillar-modified electrodes as well as the presence of rough surfaces at the top end of the nanopillars as seen in the SEM image. A similar observation for electrodes of rough surfaces was made by others where they attributed it to the presence of a large number of edges in the rough surfaces leading to more defects in the SAM structures [21, 22].

3.3.4   Calibration for Detection Sensitivity

We now take a close look how these two chemical modification techniques, using SAM versus using conducting polymers, fare in the overall glucose detection performance. Figure 3.16 shows the measured amperometric current responses of the SAM treated and GOx functionalized electrodes as drops of glucose are added sequentially in a similarly manner as discussed in Sect. 3.2.6. Interestingly, the current level for the MPA treated electrodes is much higher than the MUA treated electrodes. For calibrating detection sensitivity, the measured current at each glucose concentration is first taken and plotted against the corresponding cumulative glucose concentration. Then the detection sensitivity is determined by the slope of each calibration plot (evaluated through a linear regression analysis) normalized by the geometric area of the corresponding electrode. The inset in Fig. 3.16 shows the variation of the measured amperometric current with glucose concentration for the two types of electrodes. The detection sensitivity is found to be 2.68 and $0.09 \,\mu A \cdot mM^{-1} \cdot cm^{-2}$ for the MPA and the MUA treated electrodes, respectively.

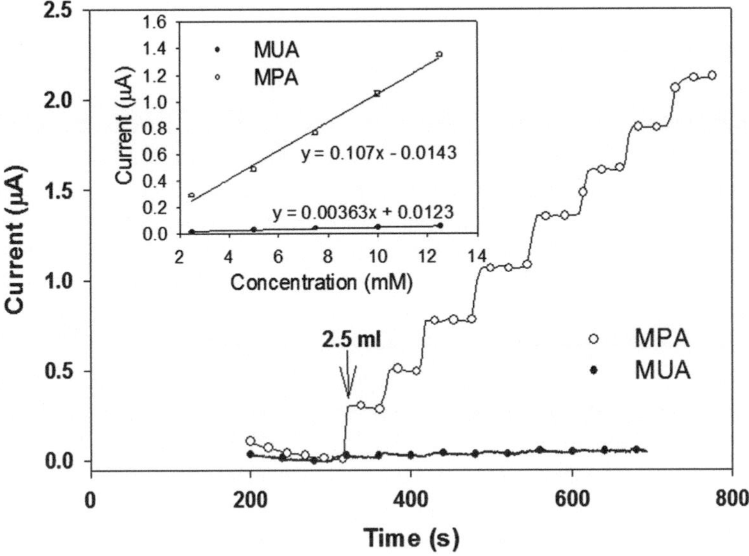

Fig. 3.16 Amperometric current measurements obtained for the MPA and MUA treated electrodes in response to glucose at various concentrations. The *inset* shows the two linear calibration curves

To put the sensitivity values of the nanopillar-modified electrodes in perspective with respect to flat electrodes, let us look at some comparisons. For the MPA and MUA treated flat (not surface modified) electrodes, the sensitivity values are found to be 0.47 and $0.052 \,\mu A \cdot mM^{-1} \cdot cm^{-2}$, respectively. For the MPA treated electrodes, the presence of the nanopillars causes an approximately sixfold increase in detection sensitivity. Although a sixfold increase is high, it is not higher enough with respect to the 45-fold increase in the surface area of these surface modified electrodes. This disparity is certainly related to the increased amount of blockage for electron transfer from the SAM molecules covering the surface of the modified electrodes, but it may also suggest that the amount of GOx functionalized onto the modified electrodes is not proportional to the available SAM surface as in the case with flat electrodes. In comparison with the nanopillar-modified electrodes chemically modified via the conducting polymer approach, it appears that the conducting polymer approach clearly wins the competition. Recall that the detection sensitivity for the electrodes functionalized using conducting polymers is approximately $36 \,\mu A \cdot mM^{-1} \cdot cm^{-2}$ (see Sect. 3.2.6). It is about 12 times higher than that for the MPA treated electrodes and about 400 times higher than that for the MUA treated electrodes. Even after accounting for the difference in their roughness factors, 60 for the conducting polymer cases and 45 for the SAM cases by applying a knock-down factor of 0.75 (45/60), the difference between the conducting polymer and SAM cases, i.e., ninefold difference, is still quite significant.

3.4 SAM Based Surface Modification for Affinity-Type Biosensors

Although we have shown in the previous sections that for an enzymatic glucose biosensor, chemical modification of electrode surfaces using conducting polymers fares much better than that using SAM, the SAM approach is still indispensable in many situations. A case in point: in an affinity type biosensor, sensitive molecules are used to couple or bind with the target molecules. Unlike an enzyme, which may still be active even it is trapped inside another molecule, the functional groups of the affinity molecules need to be freely exposed for binding with targets. In this situation, the SAM approach becomes necessary and useful.

In this section we will examine a case where nanopillar-modified electrodes are used as an affinity type biosensor. For this purpose, similar steps as discussed in previous sections are taken for fabricating the surface modified electrodes and preparing them for testing. The electrodes used here have a roughness factor of about 38 and only MUA is used as the anchoring SAM. After MUA activation (see detailed discussion in Sect. 3.3.1), the electrodes are placed in PBS solution containing $300\,\mu L$ of avidin at $200\,\mu g/mL$ for 2 h at room temperature for avidin immobilization. With further rinsing with PBS (0.01 M, pH 7.4), the avidin-functionalized electrodes are tested in reaction with biotin of different concentrations from 1 to $50\,ng/mL$ in PBS solution. Figure 3.17 shows a schematic illustration depicting the stepwise procedure in which surface molecular adsorption and affinity coupling occur at the surface of the electrodes.

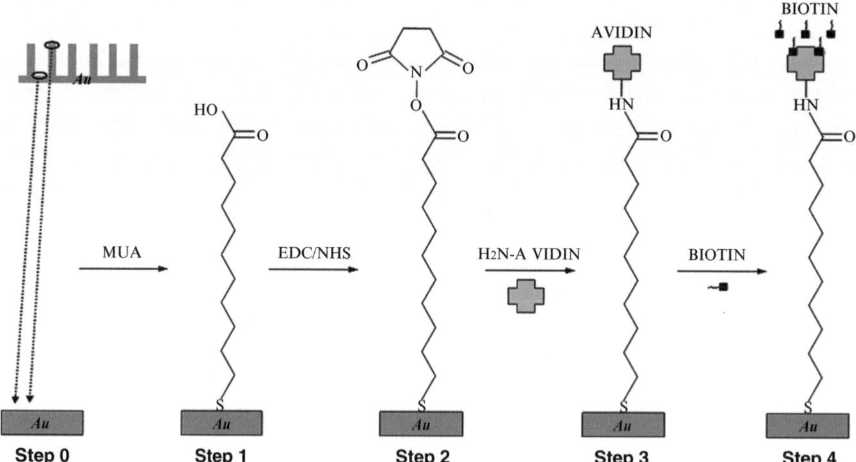

Fig. 3.17 Schematic illustration of a sequential procedure used to modify the surface of a gold nanopillar-modified electrode

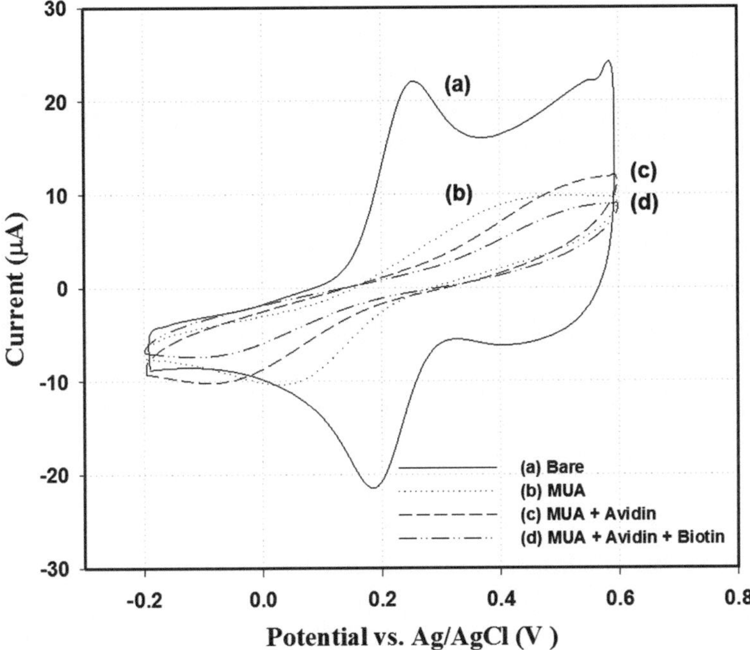

Fig. 3.18 CV measurements for examining the electrode/electrolyte interface after stepwise molecular adsorption

3.4.1 Surface Adsorption of MUA, Avidin and Biotin

Figure 3.18 shows the CV measurements taken during the stepwise surface modification of a nanopillar-modified electrode with MUA, avidin, and biotin. Clearly, in the beginning the bare electrode exhibits almost identical behavior as we see in Fig. 3.12 with a peak-shaped CV curve and peak-to-peak separation (ΔE_p) of approximately 60 mV for the oxidation and reduction of $Fe(CN)_6^{3-/4-}$, indicating that a reversible and efficient redox event is going on across the electrode/electrolyte interface. After the MUA layer is adsorbed onto the electrode surface, the peak-shaped CV curve exhibits much reduced peak currents and an increased peak-to-peak separation, suggesting blockage for electron transfer due to MUA adsorption. With more layers of molecular adsorption (i.e., avidin and biotin), the CV curve shows no obvious redox peaks, indicating that the adsorption of molecules to the electrode surface has significantly lowered the electron transfer rate such that the electrode process is no longer controlled by a diffusion process but by the rate of electron transfer.

From the EIS measurements shown in Fig. 3.19, it is seen that the Nyquist plot for the bare electrode (see Inset-1) is nearly a straight line, a typical characteristic of a diffusion-controlled electrode process. As molecules are added to the

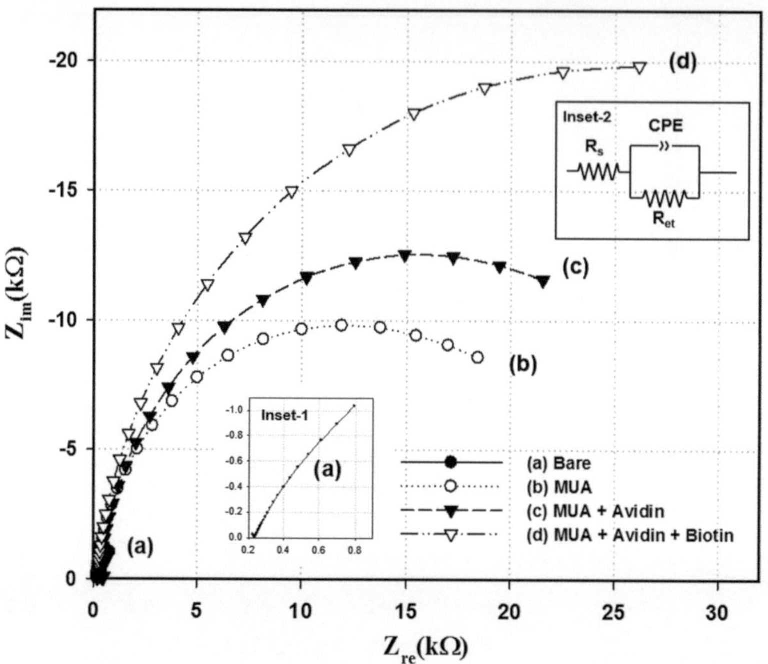

Fig. 3.19 Nyquist plots for examining the solution and electron-transfer resistance. *Inset-1*: A close view of the trace for the untreated bare case. *Inset-2*: A Randles equivalent circuit

surface, semicircular Nyquist impedance spectra are observed. The diameter of these Nyquist semicircles increases with sequential adsorption of MUA, avidin, and biotin, pointing to the fact that the electron-transfer resistance at the electrode/electrolyte interface increases as more layers of molecules are added to the surface. Moreover, these Nyquist plots do not possess a linear part at low frequency, thus confirming that after the adsorption of various molecules the electrode process is no longer a diffusion-controlled process but a kinetics-controlled one.

In comparing the CV and EIS measurements, it is seen that both measurements not only show remarkable changes due to molecular adsorption to the electrode surface but also reveal the resistive and capacitive nature of the electrode/electrolyte interface. The change in EIS measurements is more distinct and follows a clear trend in which the Nyquist diameter increases as more layers of molecules are adsorbed to the electrode surface. By contrast, the change in CV measurements shows several transitions: (1) from curve (a) to curve (b) during which the peak current decreases and the peak-to-peak separation increases; (2) from curve (b) to curve (c) where the peak-shaped CV transitions to a non-peak-shaped CV; and (3) from curve (c) to curve (d) whereas the limiting current decreases. Therefore, with nanopillar-modified electrodes, the EIS measurements are more sensitive and suitable, as compared with the CV measurements, for discriminating the slight change caused by surface adsorption of MUA, avidin, and biotin.

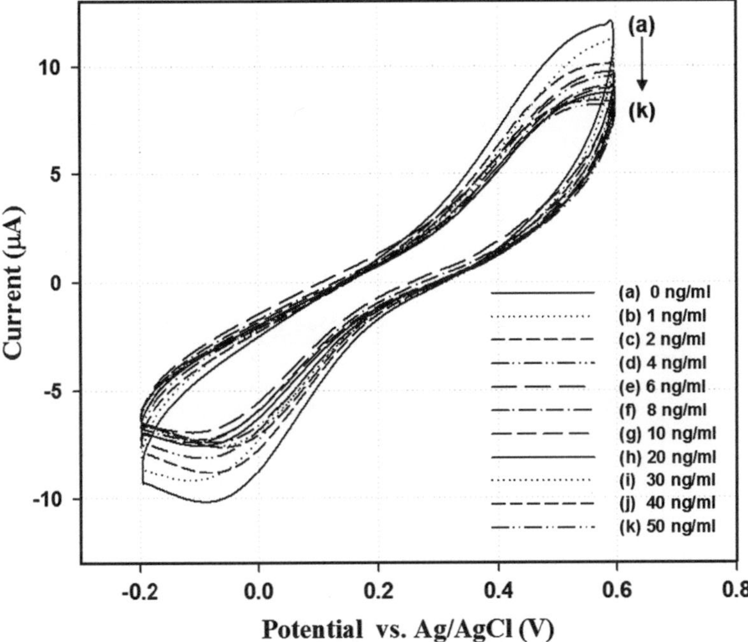

Fig. 3.20 CV for examining the electrode/electrolyte interface after the coupling of avidin with biotin at various biotin concentrations

3.4.2   *Coupling of Avidin and Biotin at Various Concentrations*

Figure 3.20 shows the CV measurements for the avidin and biotin interaction in a PBS solution containing the redox couple of $[Fe(CN)_6]^{4-/3-}$ at various biotin concentrations (i.e., 1, 2, 4, 6, 8, 10, 20, 30, 40, and 50 ng/mL). Unlike in the bare electrode case, these CV curves are not peak-shaped, indicating that the coupling of biotin with avidin at the electrode surface has caused further reduction in electron transfer. Moreover, the limiting current of these CV curves decreases with increasing biotin concentration, suggesting that more biotin is coupled with avidin as biotin concentration increases. The change in the CV measurements, however, is hard to discern, especially at a higher biotin concentration. For instance, at a biotin concentration higher than 8 ng/mL, the CV curves seem to stack on top of each other, thus making it very difficult, if not impossible, to discriminate the concentration-dependent responses.

Figure 3.21 shows the corresponding EIS measurements for the same avidin-biotin coupling experiments. In all cases, semicircular Nyquist impedance spectra are observed and these Nyquist plots do not possess a linear part at low frequency, confirming that the electrode process is not diffusion-controlled but rather kinetics-limited. As the biotin concentration increases, a distinct change in the Nyquist semicircle is seen: the higher the biotin concentration, the larger the diameter of the

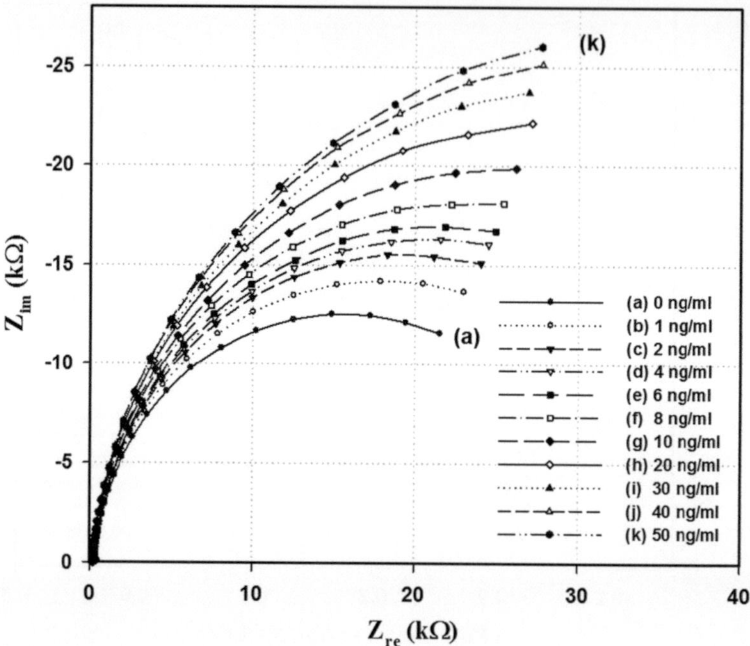

Fig. 3.21 EIS measurements (Nyquist plots) for examining the electrode/electrolyte interface after the coupling of avidin with biotin at various biotin concentrations

Nyquist semicircle. Since a larger Nyquist semicircle represents a higher electron-transfer resistance, this result indicates blockage for electron transfer at the electrode surface due to the coupling of avidin and biotin.

As discussed in Sect. 3.3.2, the resistive and capacitive behavior of an electrode/electrolyte interface can be quantitatively evaluated by using a Randles equivalent circuit (shown again as inset-2 in Fig. 3.19). The physical origin of the CPE has been widely discussed in the literature. The impedance of the CPE can be written as

$$Z_{CPE} = \frac{1}{T(i\omega)^P} \tag{3.4}$$

where i is the imaginary symbol of a complex variable, ω is the angular frequency, P is a constant ($0 \leq P \leq 1$), and T measures the inverse impedance contribution of the CPE. The variable T takes the unit of $F(rad/s)^{1-P}$, which varies with the value of P. When $P = 1$, T is equivalent to an ideal capacitor thus taking the unit of capacitance (i.e., F); when $P = 0$, T is equivalent to an inverse resistor thus taking the unit of inverse resistance (i.e., $1/\Omega$; note that $F \times (rad/s) = 1/\Omega$).

Table 3.3 Evaluation of molecular surface adsorption

Surface adsorption step	R_{et} (kΩ)	R_s (Ω)	T (μF(rad/s)$^{1-p}$)	P
Step 1: MUA	21.950.16	238.82.9	3.7290.067	0.93
Step 3: Avidin	26.620.20	235.92.8	3.5320.064	0.95
Step 4: Biotin (ng/mL)				
1	31.81 $\pm$ 0.24	242.7 $\pm$ 3.1	3.214 $\pm$ 0.036	0.93
2	34.59 $\pm$ 0.26	248.7 $\pm$ 2.9	3.115 $\pm$ 0.024	0.92
4	36.20 $\pm$ 0.30	246.9 $\pm$ 1.7	3.010 $\pm$ 0.035	0.92
6	37.56 $\pm$ 0.28	249.1 $\pm$ 3.7	2.901 $\pm$ 0.017	0.92
8	40.66 $\pm$ 0.27	253.4 $\pm$ 3.9	2.824 $\pm$ 0.016	0.92
10	43.83 $\pm$ 0.36	249.7 $\pm$ 2.3	2.719 $\pm$ 0.022	0.92
20	47.87 $\pm$ 0.32	258.1 $\pm$ 3.1	2.659 $\pm$ 0.016	0.93
30	49.86 $\pm$ 0.37	251.8 $\pm$ 3.0	2.614 $\pm$ 0.014	0.93
40	51.33 $\pm$ 0.19	256.3 $\pm$ 3.1	2.581 $\pm$ 0.014	0.92
50	53.55 $\pm$ 0.16	255.1 $\pm$ 2.2	2.501 $\pm$ 0.014	0.92

From the statistical fit of the Randles equivalent circuit to the obtained Nyquist plots, the corresponding elemental parameters are determined as listed in Table 3.3. After MUA and avidin adsorption, (steps 1 and 3), the obtained P value is 0.93 and 0.95, respectively, very close to 1 (an ideal capacitive response), suggesting that the CPE resembles more as a capacitor than an inverse resistor. The R_s, the solution resistance which is not expected to vary much when layers of molecules are adsorbed onto the electrode surface, changes only slightly. The R_{et} value, the electron-transfer resistance, increases as layers of molecules are sequentially adsorbed to the electrode surface.

For the beginning bare electrode, R_{et} is about 7.789 kΩ. After the MUA layer, it increases to 21.95 kΩ, and it reaches 26.62 kΩ after avidin adsorption. The T value, however, decreases from 3.729 μF(rad/s)$^{1-P}$ for the MUA adsorbed electrode to 3.532 μF(rad/s)$^{1-P}$ after avidin adsorption. The coupling of biotin with avidin has caused the R_{et} value to further increases from 31.81 kΩ at 1 ng/mL of biotin to 53.55 kΩ at 50 ng/mL of biotin and the T value to decrease from 3.214 μF(rad/s)$^{1-P}$ at 1 ng/mL of biotin to 2.501 μF(rad/s)$^{1-P}$ at 50 ng/mL of biotin with a P value varying within 0.92–0.93.

3.4.3 Calibration for Detection Sensitivity for Avidin–Biotin Interaction

Since all the P values (within the range of 0.92–0.95) are very close to 1, the CPE in the equivalent circuit is expected to contribute more as a capacitor rather than an inverse resistor, thus one can assume $C_{CPE} = T$. To characterize the detection sensitivity of the nanopillar-modified electrodes in discriminating different

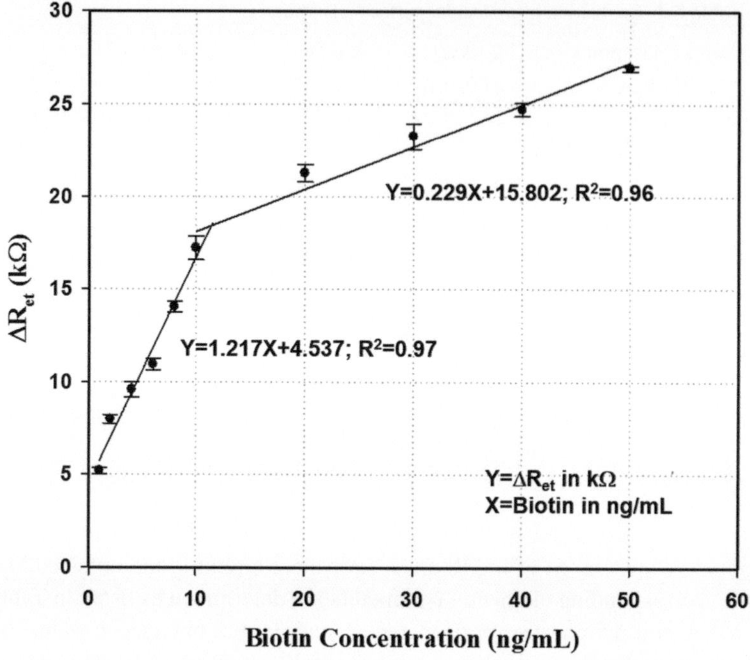

Fig. 3.22 Variation of ΔR_{et} with biotin concentration along with calibration curves

concentrations of biotin, the R_{et} and $1/T$ values obtained at each biotin concentration (Step 4) are offset by their corresponding values after avidin adsorption (Step 3), namely, taking the values of $\Delta R_{et} = R_{et(3)} - R_{et(4)}$ and $1/\Delta C = 1/T_{(3)} - 1/T_{(4)}$ as these two differential values capture both resistive and capacitive changes caused by the avidin/biotin binding. Knowing the differential ΔR_{et} and $1/\Delta C$ values at each biotin concentration, one can then construct calibration plots. Figure 3.22 shows the calibration curve of ΔR_{et} against biotin concentration. Clearly, two distinct linear segments exist. Within the range of 1–10 ng/mL, the calibration line has a slope of $1.217\,k\Omega \cdot ng^{-1} \cdot mL$, and within the range of 10 to 50 ng/mL the calibration line has a slope of $0.229\,k\Omega \cdot ng^{-1} \cdot mL$. By considering the geometric surface area ($1.2 \times 1.2\,mm^2$) of the electrode, a detection sensitivity for ΔR_{et} is determined as 845.1 and $159.0\,\Omega \cdot ng^{-1} \cdot mL \cdot mm^{-2}$, respectively, for the two linear segments. Figure 3.23 shows the calibration curve of $1/\Delta C$ against biotin concentration. Again two linear segments exist: the calibration line has a slope of $6.043\,(mF \cdot ng)^{-1} \cdot mL$ within 1–10 ng/mL of biotin and $0.750\,(mF \cdot ng)^{-1} \cdot mL$ within 10–50 ng/mL. By normalizing these values with respect to the geometric surface area of the electrode, detection sensitivity for $1/\Delta C$ is calculated as 4.196 and $0.521\,(mF \cdot ng)^{-1} \cdot mL \cdot mm^{-2}$, respectively, for the two segments.

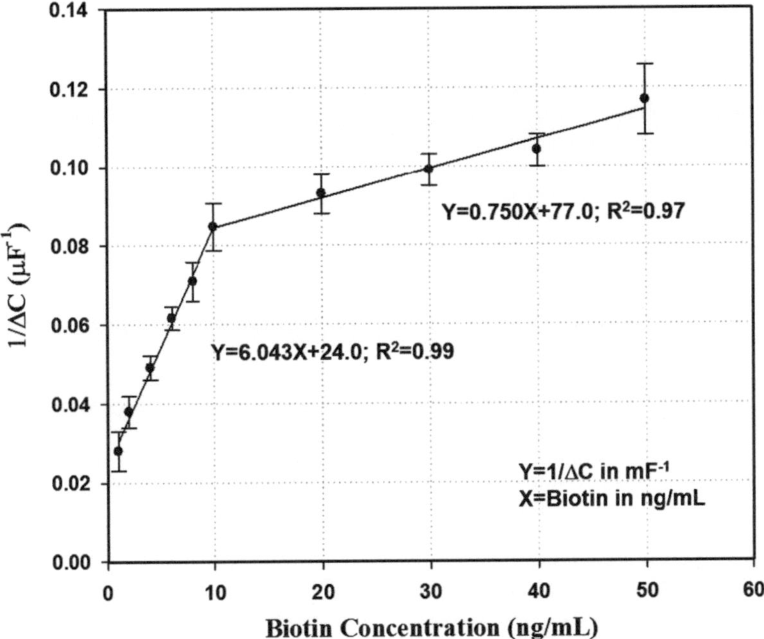

Fig. 3.23 Variation of $1/\Delta C$ with biotin concentration along with calibration curves. Note that in this case the Y fit curve has units of $(mF)^{-1}$ instead of $(\mu F)^{-1}$

References

1. Ramanaviius, A., Ramanaviien, A., Malinauskas, A.: Electrochemical sensors based on conducting polymerpolypyrrole. Electrochim. Acta **51**(27), 6025–6037 (2006)
2. Wang, J., Nosang, M., Minhee, Y., Harold, M.: Glucose oxidase entrapped in polypyrrole on high-surface-area Pt electrodes: a model platform for sensitive electroenzymatic biosensors. J. Electroanal. Chem. **575**:139–146 (2005)
3. Chen, C., Jiang, Y., Kan, J.: A noninterference polypyrrole glucose biosensor. Biosens. Bioelectron. **22**(5), 639–643 (2006)
4. Uang, Y.M., Chow, T.C.: Criteria for designing a polypyrrole glucose biosensor by galvanostatic electropolymerization. Electroanalysis **14**, 1564–1570 (2002)
5. Ramanaviius, A., Kauait. A., Ramanaviien, A.: Polypyrrole-coated glucose oxidase nanoparticles for biosensor design. Sensors Actuators B **111–112**, 532–539 (2005)
6. Wang, J., Musameh, M.: Carbon-nanotubes doped polypyrrole glucose biosensor. Anal. Chim. Acta **539**(1–2), 209–213 (2005)
7. Yang, M., Qu, F., Lu, Y., He, Y., Shen, G., Yu, R.: Platinum nanowire nanoelectrode array for the fabrication of biosensors. Biomaterials **27**(35), 5944–5950 (2006)
8. Gao, M., Gordon, L.D.: Biosensors based on aligned carbon nanotubes coated with inherently conducting polymers. Electroanalysis **15**, 1089–1094 (2001)
9. Fortier, G., Brassard, E., Blanger, D.: Optimization of a polypyrrole glucose oxidase biosensor. Biosen. Bioelectron. **5**(6), 473–490 (1990)
10. Anandan, V., Yang, X., Kim, E., Rao, Y., Zhang, G.: Role of reaction kinetics and mass transport in glucose sensing with nanopillar array electrodes. J. Biol. Eng. **1**(1), 5 (2007)

11. Lambrechts, M., Sansen, W.M.C.: Biosensors: Microelectrochemical Devices. CRC, Boca Raton (1992)
12. Lee, S.J., Anandan, V., Zhang, G.: Electrochemical fabrication and evaluation of highly sensitive nanorod-modified electrodes for a biotin/avidin system. Biosens. Bioelectron. **23**(7), 1117–1124 (2008)
13. Chaki, N.K., Vijayamohanan, K.: Self-assembled monolayers as a tunable platform for biosensor applications. Biosens. Bioelectron. **17**(1–2), 1–12 (2002)
14. Ding, S.J., Chang, B.W., Wu, C.C., Lai, M.F., Chang, H.C.: Impedance spectral studies of self-assembly of alkanethiols with different chain lengths using different immobilization strategies on Au electrodes. Anal. Chim. Acta. **554**(1–2), 43–51 (2005)
15. Campuzano, S., Galvez, R., Pedrero, M., de Villena, F.J.M., Pingarron, J.M.: Preparation, characterization and application of alkanethiol self-assembled monolayers modified with tetrathiafulvalene and glucose oxidase at a gold disk electrode. J. Electroanal. Chem. **526**(1–2), 92–100 (2002)
16. Losic, D., Gooding, J.J., Shapter, J.G., Hibbert, D.B., Short, K.: The influence of the underlying gold substrate on glucose oxidase electrodes fabricated using self-assembled monolayer. Electroanalysis **13**(17), 1385–1393 (2001)
17. Walczak, M.M., Popenoe, D.D., Deinhammer, R.S., Lamp, B.D., Chung, C., Porter, M.D.: Reductive desorption of alkanethiolate monolayers at gold: a measure of surface coverage. Langmuir **7**(11), 2687–2693 (1991)
18. Sawaguchi, T., Sato, Y., Mizutani, F.: In situ STM imaging of individual molecules in two-component self-assembled monolayers of 3-mercaptopropionic acid and 1-decanethiol on Au(111). J. Electroanal. Chem. **496**(1–2), 50–60 (2001)
19. Widrig, C.A., Chung, C., Porter, M.D.: The electrochemical desorption of n-alkanethiol monolayers from polycrystalline Au and Ag electrodes. J. Electroanal. Chem. **310**, 335–359 (1991)
20. Imabayashi, S., Iida, M., Hobara, D., Feng, Z.Q., Niki, K., Kakiuchi, T.: Reductive desorption of carboxylic-acid-terminated alkanethiol monolayers from Au(111) surfaces J. Electroanal. Chem. **428**, 33–38 (1997)
21. Losic, D., Shapter, J.G., Gooding, J.J.: Influence of surface topography on alkanethiol SAMs assembled from solution and by microcontact printing. Langmuir **17**(11), 3307–3316 (2001)
22. More, S.D., Graaf, H., Baune, M., Wang, C., Urisu, T.: Influence of substrate roughness on the formation of aliphatic self-assembled monolayers (SAMs) on silicon (100). Jpn. J. Appl. Phys. **41**, 4390 (2002)

Chapter 4
Adding Nanoparticles in Chemical Modification

Abstract While optimizing the functionalization process for electrodes with morphologically modified surface for improving detection sensitivity is important, sensitivity is not the only thing needing improvements in a biosensor. This chapter discusses the use of gold nanoparticles (GNPs) for improving the stability of enzymes (i.e., glucose oxidase) for the purpose of prolonging the functionality of glucose biosensors. It begins with a brief list of several common techniques for making GNPs with different sizes. It then discusses the effect of adding nanoparticles with different sizes in the functionalization process of electrodes with nanopillar-modified surface via the polypyrrole/GOx method. In elucidating the coupling characteristics of these nanoparticles with enzymes and conducting polymers, colloidal characterization in terms of zeta potential, UV–Vis absorbance spectroscopy and UV–Vis fluorescence spectroscopy are taken at various molecule–particle coupling stages. At the end, this chapter presents the sensing performances of electrodes with nanopillar-modified surface and functionalized with addition of nanoparticles over a time period up to 120 days.

4.1 The Possibility of Using Nanoparticles to Improve Enzyme Stability

In Chap. 3 we noted that nanopillar-modified electrodes functionalized using polypyrrol/Gox fare better than those functionalized using SAM/Gox in terms of glucose detection sensitivity. For many biosensors, however, sensitivity is just one of the many important attributes. For amperometric based glucose sensor for example, the stability of enzymes is also a very important issue. Typically, GOx enzymes lose their activity in days or weeks. It is therefore very important to improve the stability of the enzymes such that a biosensor will be functional over a prolonged period of time.

Gold nanoparticles (GNPs) have been extensively used in electrochemical biosensors to immobilize enzymes to electrode surfaces, mediate electrochemical reactions as redox catalysts, and amplify signals in electrochemical immunoassays [1]. They have also been used in biosensors to enhance the activity of glucose oxidase [2] and improve the stability of glucose oxidase at higher temperatures [3]. Proteins in general are known to show enhanced stability when adsorbed on

© Springer International Publishing Switzerland 2015   57
G. Zhang, *Nanoscale Surface Modification for Enhanced Biosensing*,
DOI 10.1007/978-3-319-17479-2_4

carbon nanoparticles or nanotubes [4]. These observations point to the possibility of adding GNPs in electrode functionalization when the surfaces of the electrodes are morphologically modified. It is known that the size of the GNPs affects their interactions with proteins [5] and that increased glucose oxidase activity is associated with reducing the size of GNPs [6]. So the questions need to be answered are, (1) will adding GNPs improve enzyme stability in situation where the electrode surfaces are nanopillar-modified? (2) What size of GNPs is optimal? In this chapter we will answer these questions by examining the effects of GNPs of different sizes. In particular, we will focus on assessing the stability of glucose oxidase in an amperometric glucose sensor over time.

4.2   Experimental Procedures

4.2.1   Reagents and Solutions

Pyrrole, β-D-glucose, glucose oxidase (GOx), sodium citrate, potassium borohydride, (Tetrakis (hydroxymethyl) phosphonium chloride), Chitosan (100 kDa), and p-benzoquinone of analytical grade from Sigma-Aldrich (St Louis, MO, USA) are used. All solutions are prepared using deionized water. Glucose solution is allowed to mutarotate for 24 h prior to experiments.

4.2.2   GNPs Synthesis

THPC Based Reduction Method Small (2–4 nm) GNPs are prepared in aqueous solution by using the reduction of chloroauric acid with Tetrakis (hydroxymethyl) phosphonium chloride (THPC) as detailed in the literature [7]. With this method, 200 μL of THPC (12 μL/mL) and 100 μL 0.6 M NaOH is added to 20 mL of water and mixed for 15 min. Then 400 μL of 1 wt% $HAuCl_4$ is added after 15 min and is left to react for another 15 min. The final solution should be in brown color.

Citrate/KBH$_4$ Reduction This method is for slightly larger GNPs (8–10 nm). 1 mL of 1 % $HAuCl_4$ aqueous solution and 1 mL of 1 % trisodium citrate are added to 100 mL DI water under vigorous stirring for 1 min. After that 0.11 % KBH_4 solution is added after 1 min under vigorous stirring for 10 min until the solution color changes to dark red.

Chitosan Method For even larger GNPs (25–30 nm) the method discussed in [8] is used. 1 mL of 1 % $HAuCl_4$ is mixed with 100 mL of 0.05 % chitosan in 1 % acetic acid. The mixture is heated to 70 °C under constant stirring for 2 h until the solution turned red.

Sodium Citrate Reduction GNPs with sizes larger than these (e.g., 40–45 nm) can be made using the method discussed in [9] with sodium citrate as the reducing agent. In this method 20 mL of 1.0 mM $HAuCl_4$ is heated up to a boil and then a 2 mL of a 1 % solution of trisodium citrate dehydrate is quickly added. The mixture is kept heating for about 10 min until the solution turned to wine red color.

Transmission Electron Microscope (TEM) is used to examine these GNPs and quantify their sizes. In each imaging run, 50 μL aliquot of corresponding GNP solutions is taken and dropped on a copper grid. The grid is then drained dry with a piece of tissue paper and let dry overnight in a desiccator prior to TEM imaging with STEM-Hitachi HD2000.

4.2.3 Electrode Functionalization with Enzyme and GNPs

Prepared nanopillar-modified electrodes with nanopillar diameter of about 50 nm and roughness factor of 18 (see Sect. 2.3 for fabrication details) are functionalized using the polypyrrole/GOx approach along with GNPs. For that, electropolymerization of pyrrole mixed with GOx and GNPs is performed under a constant current density of 35 μA/cm^2 for 35 min in 0.1 M KCl solution containing 0.05 M pyrrole, 0.5 mg/mL GOx (at pH 7.2). The amount of GNP solution added to the mixture varies slightly from 3 to 6 μL/mL depending on the sizes of the GNPs (determined through a separate optimization study).

Electrochemical measurements (CV and EIS), UV–Vis spectra, and zeta potential are taken to evaluate the functionalization steps and elucidate how the GNP/enzyme and GNP/pyrrole couplings enhance the stability of the GOx. For CV and EIS measurements, similar experiments as those discussed in Chap. 3 are performed. For UV–Vis measurements, UV–Vis absorbance spectra are first collected (between wavelengths from 400 to 800 nm, with a resolution of 1 nm) with GNPs/GOx mixture solutions in a 24-well plate. For the solutions, the concentration of GOx is fixed at 1 mg/mL while the GNPs concentration varies from 85 to 450 μM. After that, fluorescent readings are taken with the excitation wavelength set at 270 nm and the resulting emission spectra recorded within a wavelength range from 300 to 500 nm. For these fluorescent readings, the concentration of GOx is set at 0.005 mg/mL and the GNPs concentration varies from 3 to 6 μL/mL. Zeta-potential of the GOx/GNPs mixture is measured using a cuvette containing 2 mL DI water with addition of 10 μL of a prepared GNPs stock solution first followed by 10 μL of 1 mg/mL GOx, and the shift in zeta potential is quantified.

After all these basic characterizations, the functionalized electrodes are evaluated in glucose by measuring the amperometric current responses to glucose to assess enzyme stability over time. These amperometric measurements are done in the same way as before except that the experimental duration is much longer: the measurements are taken on Day 1, Day 3, Day 12, and Day 120.

4.3   TEM Views of GNPs

Figure 4.1 shows the TEM images of the various GNPS prepared using the different methods. Note that the scale bars are different in these images. From these images, it is estimated that the average diameter of GNPs obtained is as expected: approximately 2–4 nm for the THPC method (Fig. 4.1a), 8–10 nm for the Citrate/KBH$_4$ method (Fig. 4.1b), 25–30 nm for the Chitosan method (Fig. 4.1c), and 40–45 nm for the sodium citrate method (Fig. 4.1d). Since the sodium citrate based method produces GNPS with diameters of 40–45 nm, almost the size of the interpillar spacing ($\sim$45 nm), these GNPs are used only in the zeta potential and UV–Vis experiments and not used in electrochemical evaluations and glucose detection experiments because they may encounter difficulties in entering the spaces between nanopillars.

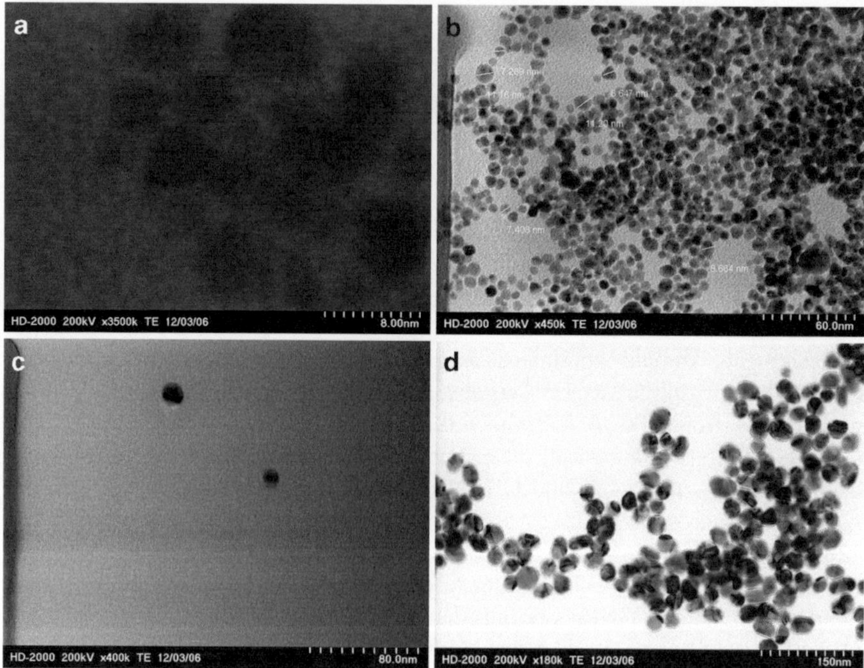

Fig. 4.1 TEM images of GNPs synthesized by using (**a**) THPC, (**b**) KBH$_4$, (**c**) Chitosan, and (**d**) Sodium Citrate reduction methods

4.4   Electrochemical Characterization of GNP Assisted Functionalization

To see whether the use of GNPs enhances or impedes the functionality of electrodes, electrochemical characterizations, i.e., CV and EIS, are performed with electrodes functionalized with polypyrrole/GOx/GNP. The CV measurements are taken in 0.1 M PBS containing 3 mM benzoquinone and 1 mM glucose to measure the oxidation and reduction peaks of the benzoquinone/hydroquinone couple in the presence of glucose and GOx.

As seen from Fig. 4.2a, the 2–4 nm case exhibits the highest oxidation peak current (0.76 mA) followed by the 8–10 nm case (0.69 mA) and the 25–30 nm case (0.34 mA). For the reduction peak, the 8–10 and 25–30 nm cases show similar peak current level (-0.74 mA) but the 2–4 nm case shows again the highest current (-0.92 mA). The fact that the 2–4 nm case generates the highest oxidation and reduction peak currents suggests that these smaller GNPs are performing better than the larger ones in soliciting the redox activities for the benzoquinone/hydroquinone couple catalyzed by GOx.

According to the cascading redox events listed in Fig. 4.3, the redox peaks of the benzoquinone/hydroquinone couple depend on the oxidation of glucose by GOx. Since the amounts of glucose and benzoquinone are kept unchanged in all these cases, the highest redox currents observed for the 2–4 nm case can only be

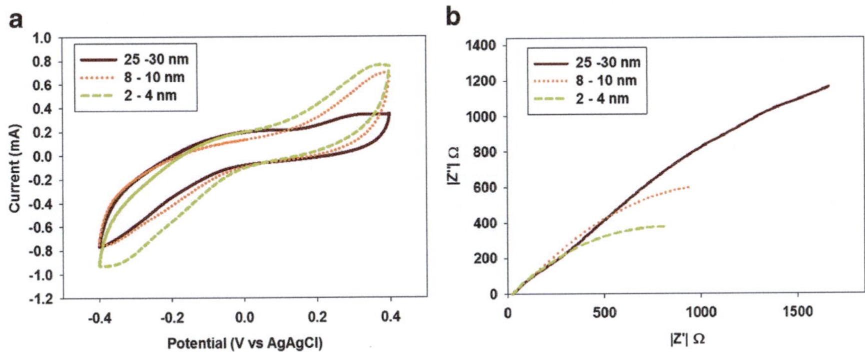

Fig. 4.2 (**a**) CV measurements for assessing the redox of hydroquinone. (**b**) The corresponding Nyquist plots

Fig. 4.3 Cascading redox activities of glucose and benzoquinone. *Thick arrows* indicate oxidation and thin ones reduction

attributed to either that a larger amount of GOx is immobilized onto the electrode or that more active sites of enzymes become available for oxidizing glucose. This fact suggests that the 2–4 nm GNPs used in electrode functionalization are indeed contributing to enhance enzyme activity. Similarly, the EIS measurements in the form of Nyquist plots given in Fig. 4.2b show that electron transfer resistance (the size of the semicircles) is much smaller for the 2–4 nm case, confirming that the 2–4 nm case has a much higher electron transfer activity.

4.5 Zeta Potential

Surface charge of these GNPs may change as they interact with enzymes, and it can be examined by measuring the change in the Zeta potential of these GNPs before and after enzyme coupling. The Zeta-potential information may shed insight into the interactions between the GNPs and enzymes. As shown in Fig. 4.4, before enzyme coupling, most of the GNPs in DI water are negatively charged except the 25–30 nm case (chitosan based GNPs) as expected because these GNPs are covered with the cationic chitosan [10]. After coupling with enzymes, the surface charge of all these GNPs increases (becomes less negative). The most change is with the 25–30 nm chitosan case and the least change is with the 2–4 nm case.

Based on the fact that at neutral pH the surface charge of glucose oxidase molecules are negative with an isoelectric point (pI) of 4.2, the results shown in Fig. 4.4 may indicate that the GNPs of 2–4 nm have better enzyme incorporation at

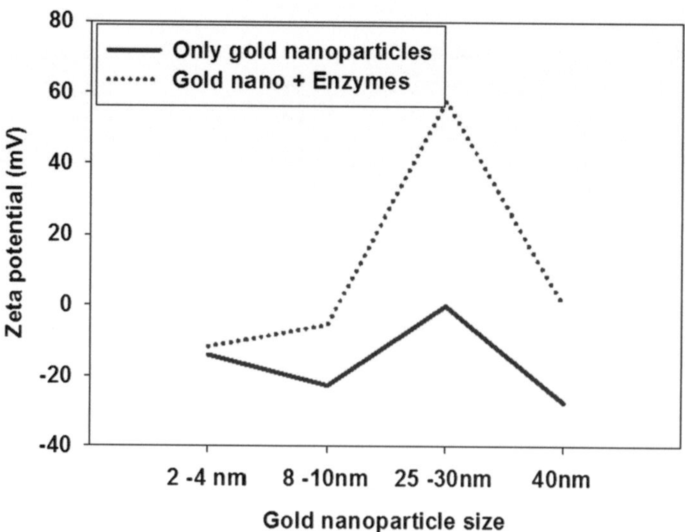

Fig. 4.4 Zeta potential of GNPs and GNP/enzyme mixture

their surfaces. The larger changes in zeta potential in the positive direction in the other three cases may point to the fact that the GNP–enzyme interactions are indeed complex, as noted from the interaction of bovine serum albumin and citrate based GNPs [11].

4.6 UV–Vis Absorbance Spectroscopy

From UV–Vis absorbance spectra one can examine the peak shift in surface plasmon resonance (SPR) of GNPs when they interact with enzymes. Figure 4.5 shows the obtained UV Vis absorbance spectra. The exact SPR peak location of GNPs is size dependent because of the mean free path of electrons is reduced as the size of the GNPs reduces [12].

From the absorbance spectra shown here, a peak around 520 nm is present in all cases. However, the 2–4 nm GNPs maybe too small to have a distinct surface plasmon absorption peak (Fig. 4.5a). After the addition of GOx a red shift (shift towards a larger wavelength) in the peak is observed for the cases of 8–10, 25–30, and 40–45 nm GNPs. This shift is larger when the nanoparticle is smaller. Note that a second peak at round 675 nm is seen for the 25–30 nm case (from the chitosan

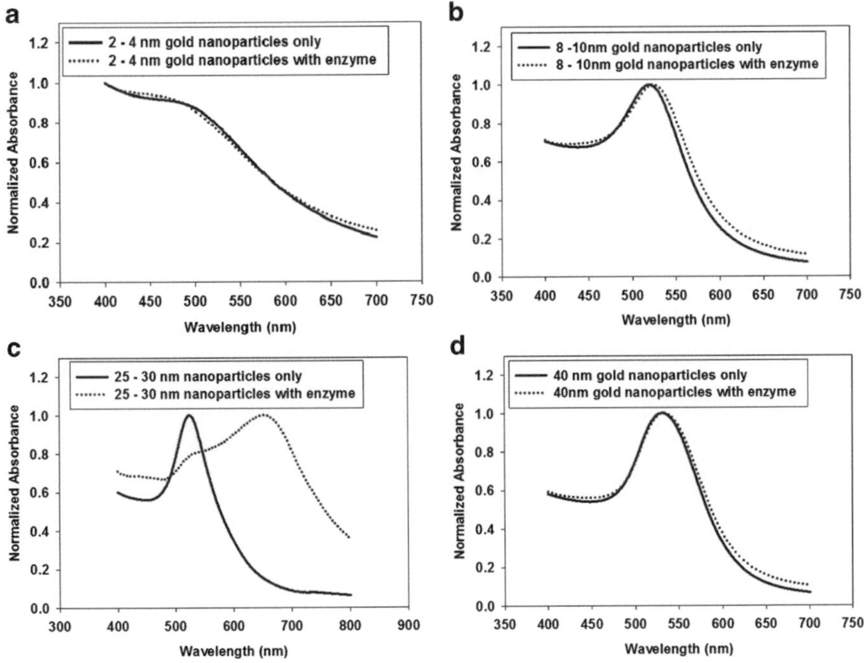

Fig. 4.5 Absorbance spectra of (**a**) 2–4 nm GNPs and GNP/GOx, (**b**) 8–10 nm GNPs and GNP/GOx, (**c**) 25–30 nm GNPs and GNP/GOx, and (**d**) 40–45 nm GNPs and GNPs/GOx

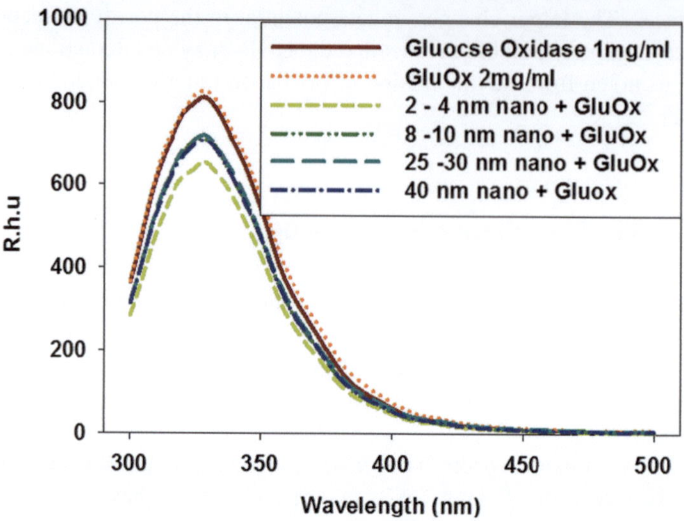

Fig. 4.6 Fluorescence spectra of enzyme and enzyme/GNPs at excitation wavelength of 270 nm

based method), and this peak is believed to be caused by the aggregation of GNPs which is known to generate peaks above 600 nm [13]. The observed red shift in the spectrum peak is the result of changes in the local refractive index around the GNPs due to enzyme binding [14], a sign of bioconjugation of GNP and GOx [15]. Overall the absorbance spectra confirm the interaction of GNPs and the enzyme molecules.

4.7 UV–Vis Fluorescence Spectroscopy

Fluorescence of tryptophan residues in proteins can be used to study the interaction between GNP and proteins [16]. Interactions of tryptophan residues with GNPs are known to quench the fluorescence of tryptophan. As seen in Fig. 4.6 the fluorescence intensity decreases with the addition of GNPs. The highest quenching occurs for the 2–4 nm case, owing possibly to the fact that the smallest GNPS can interact with GOx on multiple sites of the molecule, thereby quenching the fluorescence signal the most. These results consistently show the enhanced interaction between the smallest GNPs and GOx enzyme.

4.8 Glucose Detection and Sensitivity Calibration

Figure 4.7 shows the amperometric response of nanopillar-modified electrodes (functionalized with polypyrrole/GOx with addition of GNPs of different sizes) to stepwise additions of 2.5 mM of glucose and their corresponding sensitivity calibration curves on different days.

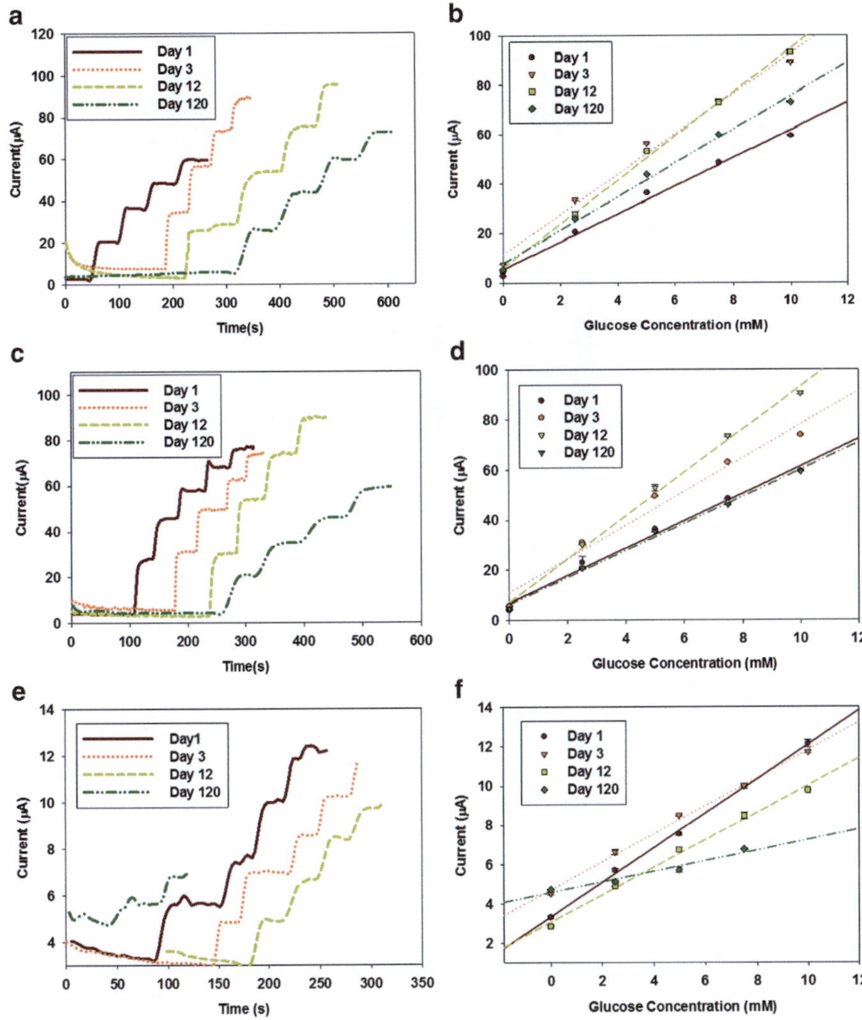

Fig. 4.7 (a) Amperometric responses to stepwise additions of glucose and the corresponding calibration curves for (**a**, **b**) 2–4 nm case, (**c**, **d**) 8–10 nm case, and (**e**, **f**) 25–30 nm case taken on Day 1, Day 3, Day 12, and Day 120

Figure 4.8 shows some bar-graph results summarizing the changes in detection sensitivity over time. The 2–4 nm case shows the best performance in terms of sensing functionality. It appears that these electrodes reach the peak performance in detection sensitivity on Day 12 ($8.87\,\mu A \cdot mM^{-1} \cdot cm^{-2}$) and lose about 23 % of its peak sensitivity on Day 120 ($6.78\,\mu A \cdot mM^{-1} \cdot cm^{-2}$). This case also shows an interesting phenomenon: the sensitivity value obtained on Day 120 is about 20 % higher than that obtained on Day 1 ($5.67\,\mu A \cdot mM^{-1} \cdot cm^{-2}$). The reason for that is not fully known. For the 8–10 nm case, the detection sensitivity is $5.44\,\mu A \cdot mM^{-1} \cdot cm^{-2}$

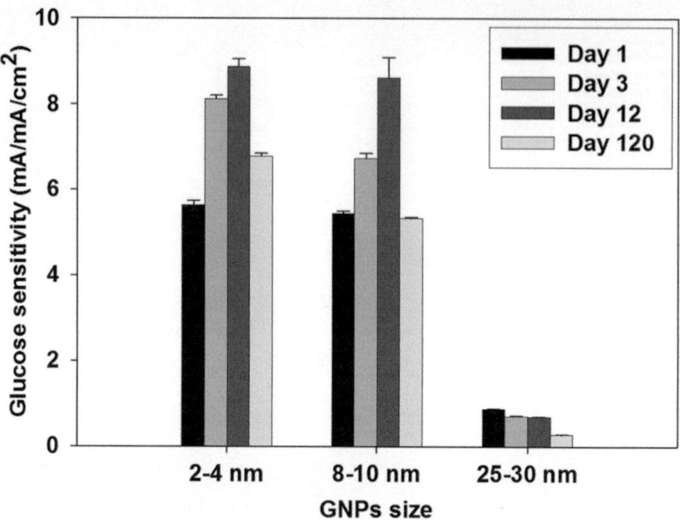

Fig. 4.8 Bar graphs of the sensitivity values for various GNPs over a time period up to 120 days

on Day 1, $6.65\,\mu\text{A}\cdot\text{mM}^{-1}\cdot\text{cm}^{-2}$ on Day 3, and $8.61\,\mu\text{A}\cdot\text{mM}^{-1}\cdot\text{cm}^{-2}$ on Day 12, with a clear increasing trend in sensitivity from Day 1 to Day 12. On Day 120, the sensitivity dropped to $5.37\,\mu\text{A}\cdot\text{mM}^{-1}\cdot\text{cm}^{-2}$, a 38 % drop from its peak performance on Day 12. The 25–30 nm case appears to be a very poor case with sensitivity values lower than $0.876\,\mu\text{A}\cdot\text{mM}^{-1}\cdot\text{cm}^{-2}$ throughout with an overall trend in losing sensitivity over time. On Day 120 almost a 70 % drop in sensitivity is seen when compared with that on Day 1.

Clearly, these results suggest that the size of the GNPs has a distinct effect on the long-term functionality of these electrodes. The smaller GNPs (2–4 nm) fare the best in improving the detection sensitivity of these electrodes and prolonging their functionality. They exhibit the least reduction in glucose detection sensitivity over a period of 120 days. This can be attributed to the enhanced interactions between the GNPS and enzyme molecules. At this moment, it is not clear what causes the increase in sensitivity in the first 12 days. It could be attributed to favorable conformation of enzymes facilitated by GNPs. If this is the case, the smallest GNPs seem to either help retain the favorable enzyme conformation over a longer period of time or prevent the enzymes from the leaching out of the polypyrrole matrix better because they increase enzymatic activity.

In comparison with the sensitivity values obtained for cases without the addition of GNPs, let us pick out the two values, 8.87 and $8.61\,\mu\text{A}\cdot\text{mM}^{-1}\cdot\text{cm}^{-2}$ for cases with 2–4 and 8–10 nm GNPs, respectively, on day 12. Recall the sensitivity values obtained in Sect. 3.2.6, i.e., $36\,\mu\text{A}\cdot\text{mM}^{-1}\cdot\text{cm}^{-2}$, by accounting for the difference in the roughness factor, 18 versus 60, we can see that the sensitivity values for the cases with GNPs are only slightly lower (note: $36\times18/60=10.80$). This slight difference could be attributed to possible measurement errors or slightly reduced amount of GOx due to the presence of GNPs.

References

1. Guo, S., Wang, E.: Synthesis and electrochemical applications of gold nanoparticles. Anal. Chim. Acta **598**(2), 181–192 (2007)
2. Pandey, P., Singh, S.P., Arya, S.K., Gupta, V., Datta, M., Singh, S.: Application of thiolated gold nanoparticles for the enhancement of glucose oxidase activity. Langmuir **23**(6), 3333–3337 (2012)
3. Li, D., He, Q., Cui, Y., Duan, L., Li, J.: Immobilization of glucose oxidase onto gold nanoparticles with enhanced thermostability. Biochem. Biophys. Res. Commun. **355**(2), 488–493 (2007)
4. Asuri, P., Karajanagi, S.S., Yang, H., Yim, T.J., Kane, R.S., Dordick, J.S.: Increasing protein stability through control of the nanoscale environment. Langmuir **22**(13), 5833–5836 (2006)
5. Jiang, X., Jiang, J., Jin, Y., Wang, E., Dong, S.: Effect of colloidal gold size on the conformational changes of adsorbed cytochrome c: probing by circular dichroism, UV–Visible, and infrared spectroscopy. Biomacromolecules **6**(1), 46–53 (2005)
6. Yeon, J.H., Park, J.K.: Microfluidic cell culture systems for cellular analysis. Biochip J. **1**(1), 17–27 (2007)
7. Duff, D.G., Baiker, A., Edwards, P.P.: A new hydrosol of gold clusters. 1. Formation and particle size. Var. Langmuir **9**, 2301–2309 (1993)
8. Huang, H., Yang, X.: Synthesis of chitosan-stabilized gold nanoparticles in the absence/presence of tripolyphosphate. Biomacromolecules **5**, 2340–2346 (2004)
9. Turkevich, J., Stevenson, C.P., Hillier, J.: A study of the nucleation and growth processes in the synthesis of colloidal gold. Discuss. Faraday Soc. **11**, 55–75 (1951)
10. Bhumkar, D., Joshi, H., Sastry, M., Pokharkar, V.: Chitosan reduced gold nanoparticles as novel carriers for transmucosal delivery of insulin. Pharm. Res. **24**(8), 1415–1426 (2007)
11. Brewer, S.H., Glomm, W.R., Johnson, M.C., Knag, M.K., Franzen, S.: Probing BSA binding to citrate-coated gold nanoparticles and surfaces. Langmuir **21**(20), 9303–9307 (2005)
12. Haiss, W., Thanh, N.T.K., Aveyard, J., Fernig, D.G.: Determination of size and concentration of gold nanoparticles from UV–Vis spectra. Anal. Chem. **79**(11), 4215–4221 (2007)
13. Shipway, A.N., Lahav, M., Gabai, R., Willner, I.: Investigations into the electrostatically induced aggregation of Au nanoparticles. Langmuir **16**(23), 8789–8795 (2000)
14. Fujiwara, K., Watarai, H., Itoh, H., Nakahama, E., Ogawa, N.: Measurement of antibody binding to protein immobilized on gold nanoparticles by localized surface plasmon spectroscopy. Anal. Bioanal. Chem. **386**(3), 639–644 (2006)
15. Shang, L., Wang, Y., Jiang, J., Dong, S.: pH-dependent protein conformational changes in albumin: gold nanoparticle bioconjugates: a spectroscopic study. Langmuir **23**(5), 2714–2721 (2007)
16. Ao, L., Gao, F., Pan, B., Cui, D., Gu, H.: Interaction between gold nanoparticles and bovine serum albumin or sheep antirabbit immunoglobulin G. Chin. J. Chem. **24**(2), 253–256 (2006)

Chapter 5
Surface Modified Electrodes in a Microfluidic Biosensor

Abstract Since the mass-transport phenomenon is quite different in an electrochemical cell setting versus a fluidic-channel setting, it is logical to ask if all the optimization and enhancement discussed in Chaps. 2, 3, and 4 are still applicable to a microfluidic biosensor. This question is better answered by examining the situation in a microfluidic setting. This chapter begins with the development of a fluidic sensor device with microchannel for fluid transport and target detection by multiple microelectrodes made of standing nanopillars. After examining the effects of flow rate, channel height and width, it presents case studies of glucose detection using the developed fluidic sensor device in which all electrodes have their surfaces morphologically modified with nanopillars and its enzyme electrode functionalized using the polypyrrole/GOx method with and without the addition of nanoparticles.

5.1 Fluidic Biosensors

For biosensor development, there is an increasing demand for sensitive detection methods integrated with microfluidics as lab-on-a-chip systems for applications such as clinical diagnostics and environmental monitoring, among others [1]. Electrochemical based biosensors are well suited for such an integration due to their good sensitivity, easiness to miniaturization, compatibility with microfabrication technologies, and low power consumption [2]. In Chaps. 3 and 4, we have seen how nanopillar-modified electrodes perform in an electrochemical-cell setting. In this chapter, we will examine how they will fare in a microfluidic setting.

Because of the presence of convective flow, the mass transport condition in a microfluidic setting is quite different from that in an electrochemical cell, we will look at the issue from two angles. First, we will examine influence of a fluidic condition on the pyrrole based functionalization process for the electrodes. Then, we will evaluate the performance of a microfluidic glucose biosensor under different flow-rate conditions.

© Springer International Publishing Switzerland 2015

G. Zhang, *Nanoscale Surface Modification for Enhanced Biosensing*,

DOI 10.1007/978-3-319-17479-2_5

5.2   Development of a Fluidic Sensor Device

The first step to develop a fluidic sensor device is to fabricate microelectrodes made of standing nanopillars on a glass slide. It is achieved by following the procedures described in Sect. 2.3. The second step is to construct a fluidic channel using polydimethylsiloxane (PDMS) and place it on top of the glass slide having the formed microelectrodes consisting of standing nanopillars. The PDMS cover containing a straight channel in dimensions of about 15 mm long, 500 μm wide, and 100 μm tall can be formed on a piece of glass with an SU-8 (an epoxy based negative photoresist) mold taking the shape of the straight channel which is developed through a conventional photolithography process. The edges of the PDMS cover is defined by a rectangular well of 40 mm long and 25 mm wide. Prior to PDMS pouring, two thin polymer tubes, one at each end of the straight channel, are placed to form the inlet and outlet of the fluidic channel. To actual form the PDMS cover, a mixture of PDMS prepolymer and its curing agent in a ratio of 10:1 is first prepared in a vacuum chamber, and then poured into the rectangular well over the SU-8 mold and left for curing for 2 h at 70 °C. The cured PDMS cover is then separated from the glass based and the SU-8 mold. After rinsing in ethanol and drying in nitrogen gas, the PDMS cover is placed over the glass substrate with its straight channel running through the parallel microelectrodes (see the Microfluidic channel part of Fig. 5.1) through a reversible bonding method. To achieve this reversible bonding, the smooth PDMS surface and the area of the glass substrate surrounding the formed electrodes are first wetted with ethanol and let dry, and then the PDMS cover is clamped to the glass substrate using a home-made acrylic clamping fixture. The reversible bond is strong enough to withstand the fluid pressure encountered during experiments.

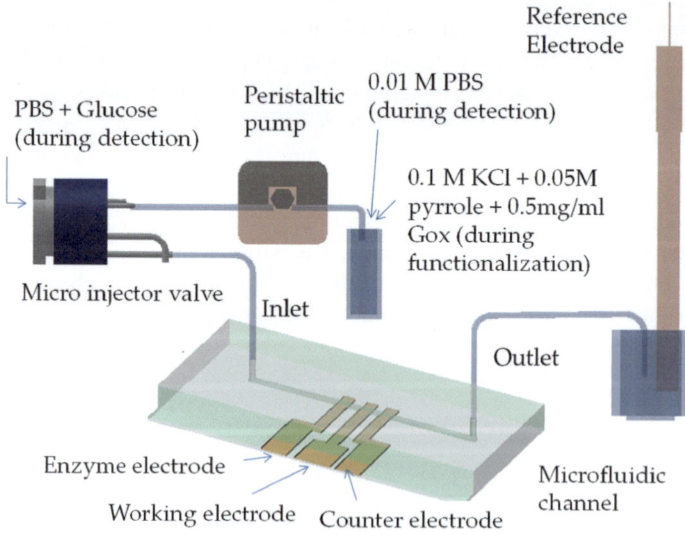

Fig. 5.1 Experimental setup developed for both electrode functionalization and amperometric glucose detection

Figure 5.1 shows the entire experimental setup consisting of a PDMS microfluidic channel having three parallel microelectrodes made of nanopillars, a micro injector valve for controlling the supply of solutions, a peristaltic pump for moving the liquid, and an Ag/AgCl reference electrode. During electrode functionalization, the peristaltic pump is connected to a reservoir containing 0.1 M KCl mixed with 0.05 M pyrrole and 0.5 mg/mL GOx while the injector valve is closed to glucose solution. During glucose detection, the pump is connected to a reservoir containing 0.01 M PBS and the injector valve is open to allow the addition fixed amounts of glucose in 0.01 M PBS.

5.3   Electrode Functionalization

In the microfluidic channel, of the three parallel electrodes the one close to the inlet (see Fig. 5.2) is functionalized with enzymes (so this one is called enzyme electrode), the middle one is used as working electrode, and the one close to the outlet is used as a counter electrode. It is worth noting that when a single electrode is used as functionalized working electrode, its sensitivity is only about 30–40 % that of the case when separate enzyme and working electrodes are used. This can be attributed to the diffusion barrier caused by the polypyrrole layer and the quick swept-away of the electroactive species in the single electrode case. For this reason, separate enzyme and working electrodes are used here.

With the setup we have here, electropolymerization can be performed directly in a fluidic setting as well. As illustrated in Fig. 5.1, for enzyme immobilization a galvanostatic process is performed by applying a constant current density at $50\,\mu A/cm^2$ to the enzyme electrode while pumping the solution through at a rate of $5\,\mu L/min$ for about 40 min against an Ag/AgCl reference electrode. The solution is a mixture of 0.1 M KCl, 0.05 M pyrrole and 0.5 mg/mL GOx. For comparison purpose, this electro-functionalization process is applied to fluidic devices having either nanopillar-modified electrodes or flat electrodes (as a control). In the case of flat electrodes, the deposition current is at $382\,\mu A/cm^2$ for 130 s according to previously established optimal values [3, 4].

To assess the formation of polypyrrole film on nanopillars, EIS measurements before and after the electro-polymerization process are taken in 10 mM PBS containing 5 mM $[Fe(CN)_6]^{3-/4-}$ and 0.15 M KCl in a frequency range from

Fig. 5.2 A close view of the microfluidic channel

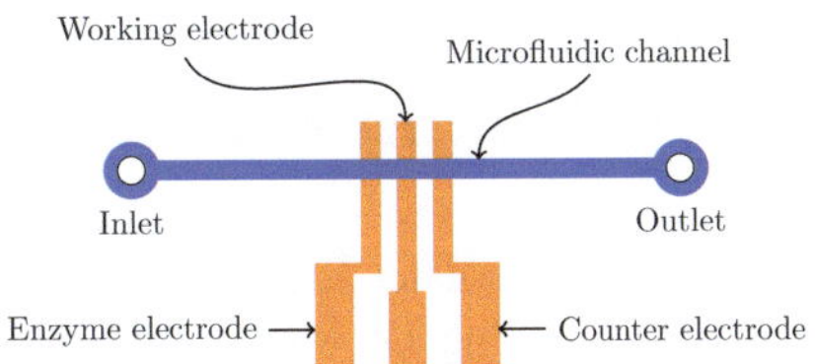

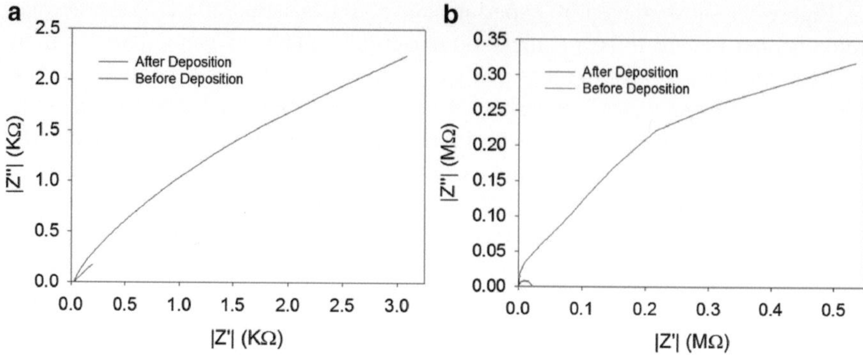

Fig. 5.3 Nyquist plots from the EIS experiment before and after electro-deposition (**a**) in an electrochemical-cell setting and (**b**) in fluidic setting

Table 5.1 Resolved R_{et} value from fitting the Randles equivalent circuit to the impedance results

Device	Deposition process	$R_{et}(k\Omega)$ before functionalization	$R_{et}(k\Omega)$ after functionalization	Increasing ratio
1	Fluidic setting	17	376	22.1
2	Electrochemical-cell setting	0.285	2.9	10.2

100 mHz to 1 kHz under an AC potential with an amplitude of 50 mV (vs. Ag/AgCl). The obtained impedance values in a fluidic setting are compared with those obtained for similar electrodes in an electrochemical-cell setting.

The EIS measurements before and after functionalization for the nanopillar electrodes in the fluidic and electrochemical-cell settings are given in Fig. 5.3a, b, respectively. The semicircular feature in the Nyquist plots is indicative of the blockage of electron transfer due to surface molecular deposition. A larger difference in the radius of the semicircles is seen for electrodes prepared in the fluidic setting as compared with those prepared in an electrochemical-cell setting. To evaluate this, the Randles circuit (see Sect. 3.3.2) fitting value for R_{et} is quantified as listed in Table 5.1. The change in the value of R_{et} before and after functionalization is about 22-fold when it is done in the fluidic setting and about tenfold in an electrochemical-cell setting. This result suggests that the functionalization performed in the fluidic setting has a better coverage of the GOx than does in the electrochemical-cell setting.

5.4   Glucose Detection

Prior to glucose detection, the functionalized fluidic channel devices are thoroughly rinsed by passing DI water followed by 0.01 M PBS at a rate of 25 μL/min. To begin the glucose detection experiments, PBS (0.01 M) containing *p*-benzoquinone

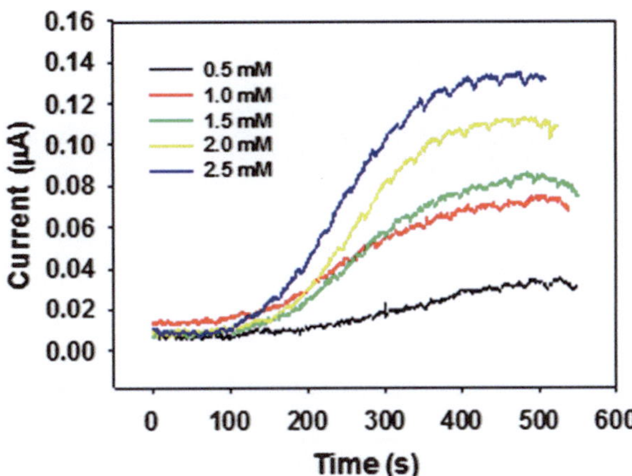

Fig. 5.4 Amperometric current responses obtained for devices with flat electrodes under various glucose concentrations

(3 mM) as a mediator is first fed through the channel device at a constant rate of 5 μL/min while the amperometric current response was measured by holding the working electrode at 0.35 V (vs Ag/AgCl). After the current response stabilized, PBS (0.01 M) containing a known concentration of glucose and the mediator is introduced to the channel at the injector. In this way, glucose is to be oxidized with the mediator reduced at the enzyme electrode (see Fig. 4.3 for the cascading reactions). The reduced mediator flows downstream and gets oxidized at the working electrode, resulting in a current response whose value depends on the glucose concentration. The experiment is repeated at various increments of glucose concentration. To compare the performance of fluidic sensors having nanopillar-modified electrodes with those having flat electrodes, the same experiment is performed with fluidic devices having flat gold electrodes.

Figure 5.4 shows the current response to glucose in a concentration range from 0.5 to 2.5 mM for a fluidic device with flat electrodes. From the calibration curve (Fig. 5.5) of the measured current versus concentration, a detection sensitivity value for the flat case is found to be 7.5 $\mu A \cdot cm^{-2} \cdot mM^{-1}$. This is much higher than other existing flat fluidic channel biosensors (e.g., 2.93 $\mu A \cdot cm^{-2} \cdot mM^{-1}$ [5]).

In addition to the sensitivity value, the current versus concentration relationship captured in Fig. 5.5 is also used to characterize the enzymatic activities with the Michaelis–Menten equation:

$$I = \frac{I_{\max} C}{K_m + C} \tag{5.1}$$

where I is the current, $I_{\max}$ is the maximum value achievable for I, C is glucose concentration, and K_m is the apparent Michaelis–Menten constant, which describes the enzymatic activity. The smaller the K_m is the more efficient the enzymatic reaction becomes. In practice, the Michaelis–Menten equation is often converted

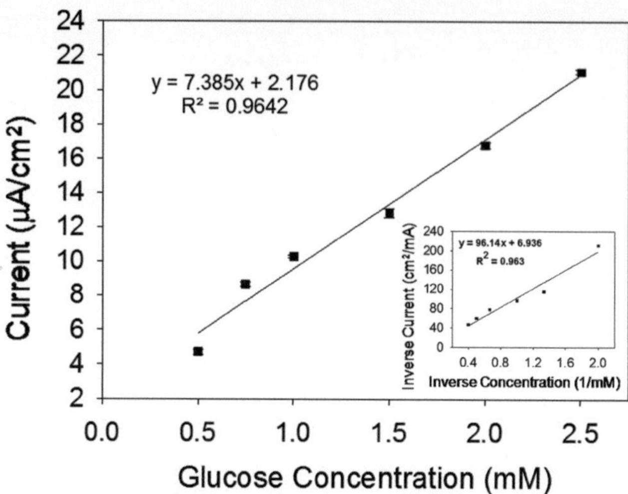

Fig. 5.5 Calibration plots for the amperometric current responses

into the Lineweaver–Burk function by taking reciprocals on both sides of the Michaelis–Menten equation:

$$\frac{1}{I} = \frac{1}{I_{\max}} + \frac{K_m}{I_{\max}}\frac{1}{C} \tag{5.2}$$

Therefore, by replotting the calibration curve as $1/I$ versus $1/C$ as in Fig. 5.5 inset, one can evaluate the kinetics of the enzymatic activity involved and thereby determine the values for K_m and $I_{\max}$. From the statistical fit to the $1/I - 1/C$ plot, the values for K_m and $I_{\max}$ are found to be 11.7 mM and 0.124 µA, respectively. This K_m value is much lower than the K_m values for glucose oxidase in a bulk solution (33 mM), suggesting that the GOx trapped inside polypyrrole matrix is highly active.

Figure 5.6 shows the current response to glucose for a fluidic device with nanopillar-modified electrodes. From the calibration plot (Fig. 5.7) the detection sensitivity value for the nanopillar case is determined to be $35.9\,\mu\text{A}\cdot\text{cm}^{-2}\cdot\text{mM}^{-1}$. This value is five times higher than that for the flat electrodes. This increase can be attributed to the enhancement of the surface area of the electrodes, allowing more surface area for enzyme immobilization and more space for efficient mass transport [6, 7]. This increase, however, is still lower than the 18 times increase in surface area. The reason for this can be attributed to that the flow condition, although helping facilitate efficient mass transport, sweep away the analytes before they are detected, indicating that the sensitivity may be further improved when the flow rate is reduced. The Lineweaver-Burk method is again used to determine the K_m and $I_{\max}$ values (Fig. 5.7) and they are 1.035 mM and 0.135 µA, respectively. Clearly, the apparent activity of enzyme is even higher in this case than in the flat case, owing to the increased surface area in nanopillar electrodes for facilitating more enzyme immobilization.

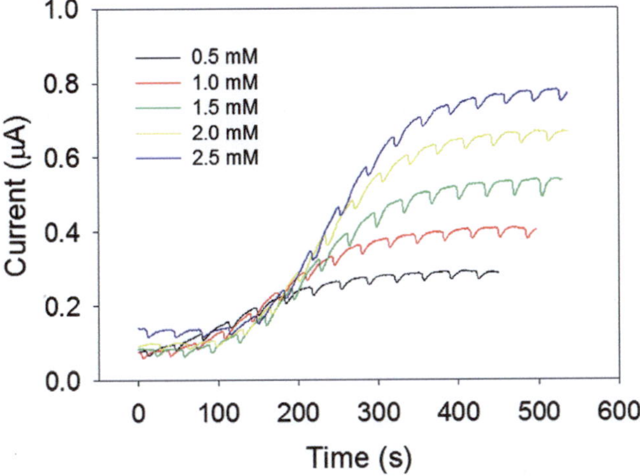

Fig. 5.6 Amperometric current responses obtained for devices with nanopillar electrodes under various glucose concentrations

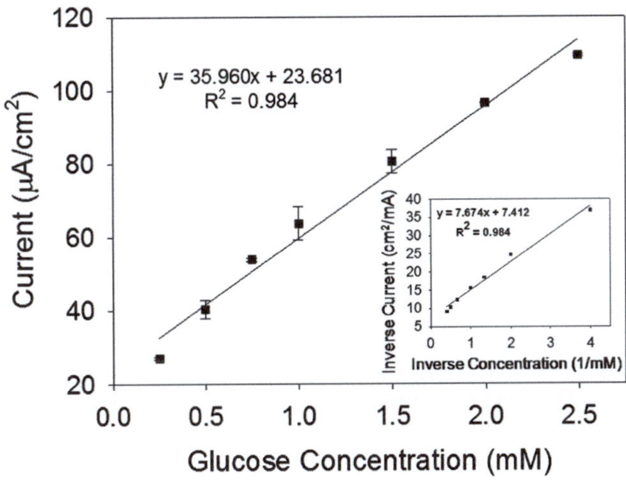

Fig. 5.7 Calibration plots for the amperometric current responses

5.5   Effect of Flow Rate, Channel Height and Width

Although the microfluidic glucose biosensor devices discussed in the previous several sections exhibit much higher detection sensitivity as compared with non-fluidic ones, it is believed that further improvements could be made through optimization. In this section, we examine the various factors including flow rate, channel height and width and height on further improving their performances.

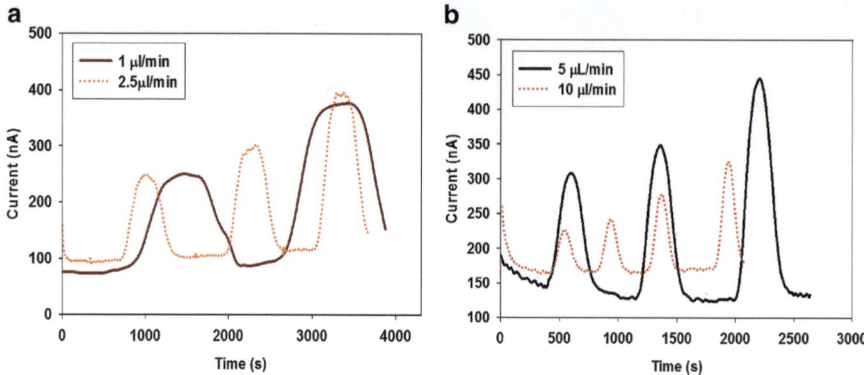

Fig. 5.8 Current responses of a microfluidic glucose biosensor having a channel of 40 μm tall and 500 μm wide obtained under various flow rates: (**a**) 1 and 2.5 μl/min and (**b**) 5 and 10 μl/min

To evaluate the effect of channel dimensions, fluidic channels of two heights and widths are made. The two heights are 20 and 40 μm and the two widths are 100 and 500 μm. Figure 5.8 shows some typical traces of the amperometric current responses of a fluidic glucose biosensor with a channel of 40 μm tall and 500 μm wide measured at four different flow rates: 1, 2.5, 5, and 10 μL/min. In each experiment the flow rate is kept constant while 20 μL of glucose at various concentrations (0.5, 1, and 1.5 mM) is introduced at the injector valve from time to time. The rise and fall seen in these plots are the result of intermittent addition of glucose at different concentrations (first rise in response to 0.5 mM glucose, second rise to 1 mM, and third rise to 1.5 mM).

It is observed from these results that the peak is wider and more flat when the rate is lower, and narrower and sharper when the rate is higher. This points to the fact that a steady state is present only when the rate is low. Lacking a steady state in the cases of higher flow rates suggests that some glucose may get washed downstream without reacting with GOx. This means that the peak current values obtained when flow rate is high may not truly represent the actual glucose concentration levels. Therefore, with a fluidic biosensor, one needs to find the right balance between the current level and the response time. When fast response is desirable, proper calibration will be required to quantify the amount of lost signals and how the measured signals are correlated to the actual levels of a target analyte (e.g., glucose in our case).

Figure 5.9 summarizes all the sensitivity values obtained for fluidic channel devices with channels of 20 and 40 μm tall and 500 μm wide under various flow rates. A clear increasing trend is visible in sensitivity values as the flow rate decreases, which is consistent with the devices having flat electrodes [8]. At 1 μL/min the sensitivity value is approximately 70 μA · cm^{-2} · mM^{-1}, about 4 times higher that at 10 μL/min ($\sim$17 μA · cm^{-2} · mM^{-1}). This value is probably the

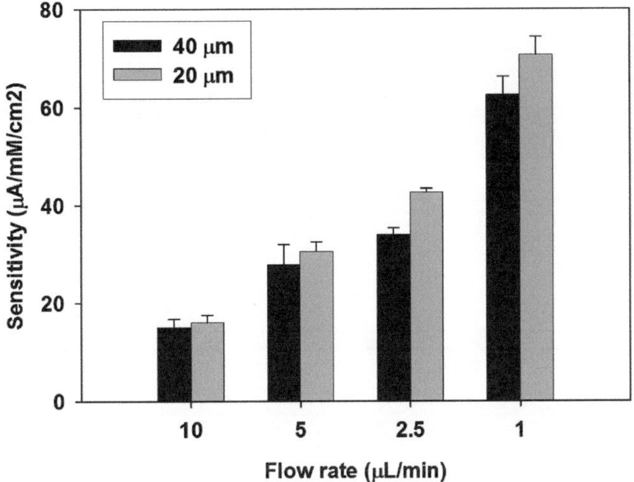

Fig. 5.9 Sensitivity values of microfluidic glucose biosensors of with channels of two different heights obtained at various flow rates

highest reported for a fluidic type of glucose biosensor. We attribute this superior performance to two facts: (1) the fluidic setting in which the efficiency of mass transport is much enhanced, and (2) the fact that separated enzyme electrode and working electrode are used with the former placed upstream and the latter downstream. Of course, to get this sensitivity, the response time will be slowed, taking a longer time for detection. On the other hand, increasing the flow rate, although it will shorten the response time, will cause significant losses of signals.

In comparison, we note that the highest sensitivity value obtained here, i.e., $70\,\mu A \cdot cm^{-2} \cdot mM^{-1}$, after accounting for the difference in their roughness factors $(70 \times 60/18 = 233)$, is more than 6 times higher than the value reported in Chap. 3, i.e., $36\,\mu A \cdot mM^{-1} \cdot cm^{-2}$. This result clearly shows the advantage of a fluidic biosensor.

In examining the effect of channel height, it is seen that the fluidic device with a taller channel $(40\,\mu m)$ shows slightly lower sensitivity values in all cases. The difference decreases as the rate increases, and in the two high-rate cases no statistical difference is found. The slight increase in sensitivity may be attributed to the fact that in a channel with a lower ceiling more glucose is pushed closer to the electrodes for reacting with GOx and generating signals. However, reducing the height further may not be beneficial because it will increase the flow rate due to fluid restriction. In assessing the effect of channel width, fluidic channel devices with channels of $20\,\mu m$ tall and 100, 200, and $500\,\mu m$ wide are prepared and tested. From these tests (data not shown), no significant change in sensitivity is noted.

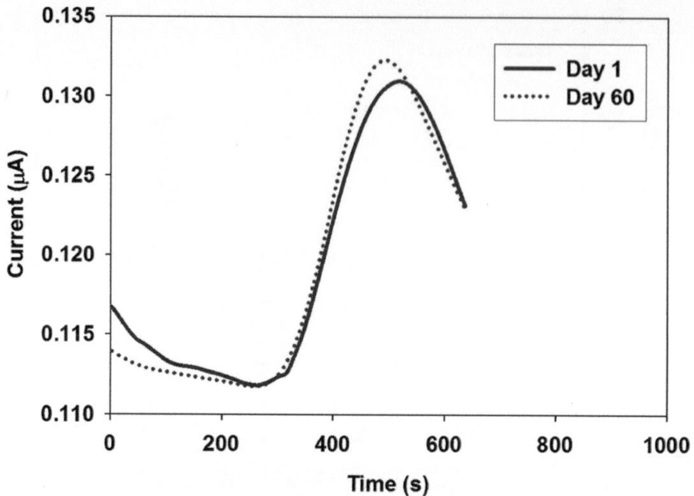

Fig. 5.10 Amperometric current response of a microfluidic glucose biosensor obtained on Day 1 and Day 60

5.6   Effect of Adding GNPs

As discussed in Chap. 4, adding small GNPs (2–4 nm) during electrode functionalization will enhance the performance of the electrodes and prolong their functionality. In this section, we will examine if the same can be achieved for a fluidic based glucose biosensor.

For this purpose, GNPs of 2–4 nm (as they are found to be most effective) are added to the solution during enzyme electrode functionalization (see discussion in Sect. 5.3). To assess these fluidic devices, glucose detection is performed over a period of 60 days. Figure 5.10 shows the amperometric current responses obtained on Day 1 and Day 60. Very surprisingly, these two measurements are almost identical, indicating that no loss of functionality and signal level after 60 days for a fluidic glucose biosensor when its microelectrodes are made of standing nanopillars and functionalized with polypyrrol/GOx with additional of GNPs of 2–4 nm.

References

1. Xu, X., Zhang, S., Chen, H., Kong, J.: Integration of electrochemistry in micro-total analysis systems for biochemical assays: recent developments. Talanta **80**, 8–18 (2009)
2. Wang, J.: Electrochemical detection for microscale analytical systems: a review. Talanta **56**, 223–231 (2002)
3. Gangadharan, R., Anandan, V., Zhang, G.: Optimizing the functionalization process for nanopillar enhanced electrodes with GOx/PPY for glucose detection. Nanotechnology **19**, 395501 (2008)

4. Uang, Y.M., Chow, T.C.: Criteria for designing a polypyrrole glucose biosensor by galvanostatic electropolymerization. Electroanalysis **14**, 1564–1570 (2002)
5. Thvenot, D., Toth, K., Durst, R., Wilson, D.: Electrochemical biosensors: recommended definitions and classification. Anal. Lett. **34**, 635 (2001)
6. Anandan, V., Rao, Y., Zhang, G.: Nanopillar arrays with superior mechanical strength and optimal spacing for high sensitivity biosensors. Proceedings of Nanotech., NSTI-Nanotech 2005, **3**, 217–220 (2005). www.nsti.org, ISBN 0-9767985-2-2
7. Anandan, V., Rao, Y., Zhang, G.: Nanopillar array structures for high performance electrochemical sensing. Int. J. Nanomed. **1**, 73–79 (2006)
8. Hashimoto, M., Upadhyay, S., Suzuki, H.: Dependence of the response of an amperometric biosensor formed in a micro flow channel on structural and conditional parameters. Biosens. Bioelectron. **21**(12), 2224–2231 (2006)

Chapter 6
Concluding Remarks

Abstract This chapter provides a summary overview of our journey in search for better ways to monitor glucose by developing nanotechnology enhanced biosensors. Through discussions of the lessons learned in this journey we hope to lay some groundwork as well as directional marks for guiding future efforts toward this pursuit, which is believed to be a long battle full of technical, social, and even cultural challenges. In addition this chapter also lays out some future perspectives for biosensor development.

6.1 What Have We Learned?

In this book we have seen that surface modification indeed plays an important role in affecting biosensor development and performance. Morphologically, adding a forest of nanopillars of only a few microns tall to a surface can increase its surface area by 10- to 100-fold. Such a remarkable increase in surface area will in turn give rises to more real estate for sensitive element immobilization and facilitate a high level of electron transfer. Of course, such benefits can only be achieved when these nanopillar structures are sufficiently robust to overcome the capillary forces encountered during solid–liquid interactions, as most biosensors will encounter aqueous environments. In this book, we not only show how aqua-robust nanopillar structures can be fabricated by using free-standing PAA template sheets, but also provide detailed discussion on how these structures can be formed on a chip or a glass slide and be further processed to form microscale electrodes of any desired patterns and shapes.

In the case studies presented, we have seen that when used as electrodes in a glucose type of biosensors, these surface-enhanced electrodes will outperform their flat counterparts with much improved detection sensitivity. Chemically, different approaches to the functionalization of these surface-enhanced electrodes will also affect their functionality. For example, when these electrodes are functionalized using the conducting polymer route, namely through electrodeposition of polypyrrole and GOx, a detection sensitivity as high as $36\,\mu A \cdot mM^{-1}cm^{-2}$ is achieved. By contrast, when electrodes are functionalized using the SAM route, namely through alkanethiols and GOx, a detection sensitivity of only $2.68\,\mu A \cdot mM^{-1}cm^{-2}$ is achieved for SAMs with a short chain length, almost 12 times lower. Although

© Springer International Publishing Switzerland 2015

G. Zhang, *Nanoscale Surface Modification for Enhanced Biosensing*,

DOI 10.1007/978-3-319-17479-2_6

the SAM approach appears to be inferior to the conductive polymer approach for enzymatic based biosensors, it may still be indispensable when comes to affinity based biosensors because the sensitive molecular probes need to be fully exposed for binding with target molecules.

Addition of gold nanoparticles in the functionalization process of these surface modified electrodes, especially small particles of 2–4 nm, results in much improved enzyme stability and subsequently prolonged sensor functionality. For example, in the case we discussed, the glucose sensor is still functional after 120 days. One cautious note is that the tests performed in this study are not only in vitro but the test solutions are PBS with glucose. This is quite different from an actual situation where whole blood samples are commonly tested, in which the presence of many proteins and other biochemical/biological species will add much complexed challenges to the biosensor. While enzyme stability is much improved with the addition of small nanoparticles and the functional duration of the resulting glucose sensors is extended, the detection sensitivity value obtained with the use of nanoparticles remains almost as high as that obtained without the use of nanoparticle in an equivalent sense, suggesting that adding smaller nanoparticles is truly advantageous.

In meeting the lab-on-a-chip challenge, we further examine the performances of these surface modified electrodes in a fluidic biosensor device. We note that the highest sensitivity achieved with a fluidic biosensor, in an equivalent sense, is more than 6 times higher than the value obtained with an electrochemical cell. This is remarkable, and it points to the superiority of a fluidic glucose biosensor in which separated enzyme electrode and working electrode are used with the former placed upstream and the latter downstream.

6.2 What is Ahead?

Combating diabetes is believed to be a long battle full of technical, social, and even cultural challenges. In summarizing the results accomplished and lessons learned in a journey searching for better ways to monitor glucose by developing nanotechnology enhanced biosensors, this book has laid some groundwork as well as directional marks for guiding future efforts for advancing this pursuit. Clearly, microfluidic based biosensors incorporated with micro- and nano-technology provide a promising platform. In the coming years it is anticipated that nanotechnology based biosensors will continue to evolve and expand their use in many areas of life sciences and biomedicine. These biosensors are expected to possess some ideal features such as high sensitivity, high specificity, fast response, low detection-limit, continuous and long-term functionality, passively operational (carries no battery, or turns the physiological metabolic events into fuel-power), and wireless operational (be able to communicate with external monitoring devices wirelessly). The integration of nanotechnology and biotechnology holds great promises for the realization of these desirable features.

It is also anticipated that future biosensor devices may become autonomous systems, capable of performing advanced diagnosis while delivering therapeutic drugs or molecules at desired locations, rate and amount in real time and on a constant and continuous basis. It will not be far stretched to assume that these devices will either be micro/nano integrated electromechanical systems, or be entirely made of organic molecular assemblies. These systems should offer reduced intrusiveness, increased patient comfort, greater fidelity of sensing results, and greater precision for site, amount and rate controllable load (drugs or therapeutic agents) delivery.

Appendix A
Detailed Processing Steps Used
for the Experiments Discussed

A.1 RCA Cleaning

Contaminants in the forms of insoluble organic residues, oxides, and ionic species on the surface of silicon wafers or glass slides need to be removed from time to time in order to obtain high-quality outcomes. The removal of these contaminates will lead to not only quality and reliable devices but also minimized contamination to the processing equipment. The RCA cleaning process is an industry standard for removing organic, oxide, and ionic contaminants from wafers and glass slides. Werner Kern developed the basic procedure in 1965 while working for RCA (Radio Corporation of America), hence the process is termed RCA cleaning.

The RCA cleaning procedure has three major steps often used sequentially:

1. Organic Clean: Removal of insoluble organic contaminants with a solution of $H_2O:H_2O_2:NH_4OH$ in 5:1:1 volume ratios. A desirable operating temperature for this solution is around 70–80 °C, thus the solution is often mixed by heating the water to a slightly higher temperature first and then add H_2O_2 and NH_4OH. Please be warned that one always adds heavy reactive compounds such as acids or bases to water, and not the other way around.
2. Oxide Strip: Removal of a thin dioxide layer where metallic contaminants may accumulate as a result of organic contamination using a diluted solution of $H_2O:HF$ in 50:1 (or even 100:1) ratio.
3. Ionic Clean: Removal of ionic and heavy metal atomic contaminants using a solution of $H_2O:H_2O_2:HCl$ in 6:1:1 volume ratios. This solution is also used at around 70–80 °C.

A warning note: HF acid is very dangerous and HF burns are particularly hazardous. An insidious aspect of HF burns is that there may not be any discomfort until long after exposure. These burns are extremely serious and may result in tissue damage. If you come in contact with HF, flush the area with cold water for 15 min and be sure to get under and around your fingernails if applicable. While rinsing,

© Springer International Publishing Switzerland 2015

G. Zhang, *Nanoscale Surface Modification for Enhanced Biosensing*,

DOI 10.1007/978-3-319-17479-2

seeking immediate assistance (you should always have a safety buddy around when working with dangerous chemicals like HF) and apply the Calcium Gluconate gel to exposed area quickly. Seek further care from a physician immediately after. Users must always strictly follow lab safety rules and requirements and consult MSDS sheets for all chemicals for their safe handling.

A.2   PVD of Multi-Layer Metal Films on Glass Slides

The following steps are performed with a PVD75 E-Beam Evaporator (Kurt J Lesker) using microscope slides 3×1 in (Corning). High purity metal pallets with matching crucibles are needed: Ti in Tungsten crucible, Au in Graphite crucible, and Al in Boron Nitride crucible (Kurt J Lesker).

A.2.1   Cleaning of Glass Slides

RCA cleaning (Note that all steps below should be completed inside a cleanroom)

1. Dip the glass substrates in 5:1:1 H_2O:H_2O_2:NH_4OH solution at around $70\,^\circ C$ for 20 min;
2. Let them cool to room temp and rinse them with DI water;
3. Dip them in highly diluted HF (1 mL HF in 50 mL H_2O) for 15 s;
4. Rinse them with DI water;
5. Blow dry with high purity N_2;
6. Bake them at $100\,^\circ C$ for 2 min;
7. Mount the cleaned glass substrates onto the sample platen using double-sided tapes.

A.2.2   Film Deposition Using E-Beam Evaporator

Loading samples and pumping down for vacuum

1. Switch on the power to the pumps;
2. Turn on cooling water partially;
3. Switch on the N_2 cylinder to supply 60 psi N_2 for all pneumatics;
4. Load all crucibles with proper materials in the order of deposition;
5. Load the sample-holder platen in the E-beam chamber;
6. Push VAC and PUMP DOWN buttons to pump down the vacuum to below 5×10^{-6} Torr.

Depositing the First Ti Layer

1. Turn on the power to the E-beam (Siemens switch)
2. Set the first layer film to Titanium via the Film Manual;
3. Turn on the main-power;
4. Turn the cooling water to full flow, wait for two or more minutes;
5. Push on the Voltage/Emission button to switch on the E-gun;
6. Switch on the sweep (Carefully adjust the sweeping range of the E-beam, if necessary);
7. Wait till the Volt reading reads about 7.33 V, start to ramp up the current slowly (manual operation);
8. For Titanium, the anticipated current for evaporation is around 0.81A (which gives a deposition rate about (dr) 0.6 Å/s; Note that this ramping up should be slow and you may open/close the Shutter to see if the evaporation point is reached you will see a change in the dr reading when reached);
9. Once the monitor shows deposition with the Shutter open, switch on the Platen Rotation (preset at 8 for 80 rpm);
10. You may need to tune the current level for a desired deposition rate (say 0.6 Å/s for Ti);
11. Monitor the thickness reading, once the thickness reading reaches the desired number, say, 50 Å($= 5$ nm), close the shutter;
12. Slowly ramp down the current to zero, switch off the sweep and the Voltage/Emission;
13. Switch off the Platen Rotation;
14. Switch off the main-power.

Depositing the Second Au Layer

1. Wait for about 10 min to let the chamber cool down;
2. Rotate the material basket to the next crucible;
3. Set the second layer film to copper via the Film Manual;
4. Turn on the main-power;
5. Push on the Voltage/Emission button to switch on the E-gun;
6. Switch on the sweep;
7. Ramp up the current slowly (manual operation);
8. For gold, the anticipated current for evaporation is around 2A (dr $= 2$ Å/s; Note that this ramping up should be slow and you may open/close the Shutter to see if the evaporation point is reached);
9. Once the monitor shows deposition with the Shutter open, switch on the Platen Rotation;
10. You may need to tune the current level for a desired deposition rate;
11. Monitor the thickness reading, once the thickness reading reaches the desired number, say, 200 Å($= 20$ nm), close the shutter;

12. Slowly ramp down the current to zero, switch off the sweep and the Voltage/Emission;
13. Switch off the Platen Rotation;
14. Switch off the main-power.

Depositing the Third Al Layer

1. Wait for about 10 min to let the chamber cool down;
2. Rotate the material basket to the next crucible;
3. Set the film setting from the Film Manual for the third layer to Aluminum;
4. Turn on the Main-Power;
5. Push on the Voltage/Emission button to switch on the E-gun;
6. Switch on the sweep;
7. Ramp up the current slowly (manual operation);
8. For Aluminum, the anticipated current for evaporation is around 0.86 A ($dr = 2$ Å/s; Note that this ramping up should be slow and you may open/close the Shutter to see if the evaporation point is reached);
9. Once the monitor shows deposition with the Shutter open, switch on the Platen Rotation;
10. You may need to tune the current level for a desired deposition rate;
11. Monitor the thickness reading;
12. Once the thickness reading reaches the desired number, say, 8 kÅ($= 800$ nm), close the shutter;
13. Slowly ramp down the current to zero, switch off the sweep and the Voltage/Emission;
14. Switch off the Platen Rotation;
15. Switch off the main power.

A.2.3   *Leaving the E-Beam in Stand-by*

1. Switch off the E-beam power supply (Siemens switch);
2. To put the system in standby, push the TURBO button, then the ROUGH button;
3. Switch off the cooling water;
4. Switch off the pump power supply (Eaton switch)
5. Close the N_2 cylinder for pneumatics.

A.2.4   *Getting the Samples Out of the E-Beam Evaporator*

1. Switch on the pump power supply (Eaton switch);
2. Open the pneumatic N_2 cylinder (60 psi) and vent N_2 cylinder (10 psi);

3. Push the VENT button;
4. Open the chamber and get the samples out.

A.3   PAA Templates Formation by Anodization

This process requires the use of a power supply (e.g., Agilent N5771, 300 V/5 A, 1,500 W), a multimeter (e.g., Keithley 2000) along with a 300 mL beaker filled with 0.3 M oxalic acid placed in the middle of a bucket with drainage. Fill the bucket with ice surrounding the beaker to keep the solution at around 5 °C. Cut a metal-film coated glass substrate into a proper size and mask the edge with Miccrostop.

A.3.1   One-Step Anodization of the E-Beam Formed Film

1. Connect the anode (positive) end of the power supply through the multimeter (for current monitoring) to the sample to be anodized, and connect the cathode (negative) end of the power supply to an Al foil (as a counter electrode), letting the sample face the Al foil;
2. Immerse the sample and the counter electrode in the oxalic solution;
3. Turn on the magnetic stirrer to keep the solution constantly stirred with a stirring bar;
4. Switch on the power supply and set the voltage output to 40 V with a current limit of 250 mA;
5. Switch on the multimeter (set on DCI for DC current monitoring);
6. Press on the OUT-ON button to start the anodization process;
7. Monitor the current reading (For 40 V anodization, there will be a surge of current in the beginning to around 20–50 mA/cm^2, then the current will stabilize around 1–2 mA/cm^2; once the anodization reaches the barrier layer, the current will drop slightly and then begin to surge—switch off the OUT-ON button when this happens; the duration usually lasts for about 20 min (for 700 nm thick Al layer) or more depending on the thickness of the Al layer to be anodized);
8. Remove the barrier layer by immersing the sample in 5 wt% phosphoric acid solution at 30 °C for about 20 min;
9. Remove the Miccrostop with acetone and rinse with DI-water.

A.3.2   Two-Step Anodization of the E-Beam Formed Film

1. Use the same setting as above;
2. The first anodization should last for only 5 min;
3. Rinse the sample in DI water;

4. Strip the formed oxide layer by dipping the sample in a 6 %wt phosphoric acid and 1.8 %wt chromic acid solution at 60 °C for 30 min;
5. Rinse the sample with DI water;
6. Place the sample back for the second anodization using the same setting until reaching the barrier layer;
7. Remove the barrier layer by immersing the sample in 5 wt% phosphoric acid solution at 30 °C for about 20 min;
8. Remove the Miccrostop with acetone and rinse with DI-water.

A.3.3   Two-Step Anodization of a High Purity Al Sheet

1. Clean a 99.9 % aluminum sheet in acetone (for degrease) and in 3.0 M NaOH solution;
2. Electro-polish the sample in 10 % perchloric acid and 90 % ethanol solution at 20 V until a mirror finish is obtained;
3. Place the sample in anodization setting (anode to the sample and cathode to an Al foil);
4. Switch on the power supply and set the voltage output to 40 V;
5. Press on the OUT-ON button to start;
6. Anodize it for about 1–2 h (slightly longer duration results a better pattern);
7. Take the sample out and rinse it;
8. Strip the oxide layer by dipping the sample in a 6 %wt phosphoric acid and 1.8 %wt chromic acid solution at 60 °C for 60 min;
9. Rinse the sample with DI water;
10. Place the sample back for the second anodization using the same setting and control the time for the desired thickness;
11. Remove the Miccrostop with acetone and rinse with DI-water;
12. Release the sample by immersing it in saturated HgCl2 till the film releases itself;
13. Rinse it with DI-water;
14. Strip the barrier layer in 5 wt% phosphoric acid solution at 30 °C for about 20 min.

A.4   Nanopillar Development Through Electrodeposition

Materials and equipment needs for this process include an electrochemical system (e.g., MultiStat System 1480; Solartron Analytical) run by a desktop computer, saturated calomel 2 MM reference electrode (Corning), platinum gauze counter electrode, heater with magnetic stir, a couple of 100 mL beakers for various solutions, gold plating solution (Orotemp 24, Technic, Inc.), and silver plating solution (ACR1025 Silver Steak, Technic, Inc.).

A.4.1   Gold Nanopillars

1. Prepare the gold plating solution (gold potassium cyanide, as purchased) in a 100 mL beaker and warm it to 65 °C;
2. Place the prepared PAA sample in the gold plating solution;
3. Use a three-electrode setting to connect the sample (working $=$ WE $+$ R2: the prepared sample; reference $=$ R1: saturated calomel; counter $=$ CT: platinum gauze);
4. Run a Galvanosatic experiment with the electrical current set at 2.5 mA/cm^2 (or 0.6 mA/cm^2 accounting for 25 % pore fraction);
5. Remove the alumina template by immersing the sample in a 100 mL beaker containing 1 M NaOH for 25 min.

A.4.2   Silver Nanopillars

1. Prepare the silver plating solution (silver potassium cyanide, as purchased) in a 100 mL beaker at room temperature;
2. Place the prepared PAA sample in the silver plating solution;
3. Use a three-electrode setting to connect the sample (working $=$ WE $+$ R2: the prepared sample; reference $=$ R1: saturated calomel; counter $=$ CT: platinum gauze);
4. Run a Galvanosatic experiment with the electrical current set at 10 mA/cm^2;
5. Remove the alumina template by immersing the sample in a 100 mL beaker containing 1 M NaOH for 25 min.

A.5   Micro-Patterning

1. Coat a micro-primer layer (HP primer) onto the sample using a spin coater;
2. Coat a Photoresist 1818 (positive photoresist) layer onto the sample with the following spin-coater setting:

    (a) 2,000 rpm—ramp up in 1 s and hold for 1 s
    (b) 4,000 rpm—ramp up in 2 s and hold for 2 s
    (c) 6,000 rpm—ramp up in 2 s, hold for 30 s and ramp down in 20 s

3. Prebake the sample at 115 °C for 4 min;
4. Transfer the desired pattern using a mask aligner (e.g., MJB3 or others) through UV exposure (15 s);
5. Postbake the sample at 100 °C for 10 min;
6. Develop the exposed photoresist in MF 319 developer solution with constant agitation for about 2 min;

7. Hard bake the sample at 100 °C for 10 min;
8. Etch the unmasked Au layer chemically in solution of $KI:I_2:H_2O$ (4:1:40 wt) at room temperature;
9. Etch the Ti layer in solution of $H_2O:HF:H_2O_2$ (20:1:1 v);
10. Strip the photoresist using photoresist remover (1165).

A.6   Fabricating Integrated Micro-Nano Electrodes

1. Glass cleaning by standard RCA procedure:

   (a) Dip the sample in an RCA organic cleaning solution ($NH_4OH:H_2O_2:H_2O$ in 1:1:5) at 75–80 °C for 10 min to remove the organic contaminants.
   (b) Immerse it in $HF:H_2O$ (1:50) at room temperature for 10 s to remove the thin oxide layer and ionic contaminants.
   (c) Dip it in $HCl:H_2O_2:H_2O$ (1:1:6) at 75–80 °C for 10 min to remove the remaining traces of metallic (ionic) contaminants.

2. Spin coating and prebake

   (a) Gold thin film on top of glass was cleaned by DI water and blew dry by Nitrogen gun;
   (b) HP prime and S1818 were spin coated successively on top of the gold with following spinning setting:

      (i)   2,000 rpm with ramp-up in 1 s and hold for 1 s
      (ii) 4,000 rpm with ramp-up in 2 s and hold for 2 s
      (iii) 6,000 rpm with ramp-up in 2 s, hold for 30 s, and ramp-down in 20 s

   (c) Prebake the sample at 95 °C for 1 min.

3. Photolithographic process

   (a) Turn on lithography power supply, and wait until the LCD shows Ready. Then press the Start button and wait until the LCD shows 194 or 195. Wait for another 5 min to let UV warm up;
   (b) Turn on the mask aligner;
   (c) Turn on the Mainframe. Make sure it is in the ST mode with soft contact on;
   (d) Before turning on the light, make sure the mark is within 4 and 5, then turn on the light;
   (e) Loosen left screw, pull out the metal plane, and flip it;
   (f) Put mask on top of metal plane, and make sure the coated side face you. Then press the vacuum mask button;
   (g) Put them into the previous position and screw tide to secure it;
   (h) Turn the lever (on the left side of the sample stage) 180° until the contact light on;

   (i) Switch another flat lever toward you to make it no contact, and use the toner in front of you to adjust the sample level to make sure there is no shadow. Then put the liver back. You can also adjust the Len to the position of your wanted structure;

  (j) Push the Exposure button allowing the exposure;

  (k) Develop the exposed sample in MF-139;

  (l) Rinse the sample in DI water, and blew dry using a Nitrogen gun;

 (m) Postbake the sample at $120\,^{\circ}\mathrm{C}$ for 5 min.

4. Metal Wet Etching After baking, the sample is etched in gold etching solution: $4\,\mathrm{g}\ KI + 1\,\mathrm{g}\ I_2 + 40\,\mathrm{mL}\ H_2O$, and in titanium etching solution $1\,\mathrm{mL}\ HF + 1\,\mathrm{mL}\ H_2O_2 + 20\,\mathrm{mL}\ H_2O$.

Index

© Springer International Publishing Switzerland 2015
G. Zhang, *Nanoscale Surface Modification for Enhanced Biosensing*,
DOI 10.1007/978-3-319-17479-2